The AXIOMS OF FLIGHT

The Rules Of Thumb For Pilots

by
James Embree

FLIGHT INFORMATION PUBLICATIONS

PUBLISHERS OF AVIATION BOOKS AND TRAINING AIDS

ST. LOUIS, MISSOURI

THE AXIOMS OF FLIGHT

The Rules Of Thumb For Pilots

ISBN 0-9601062-7-8

ABOUT THE AUTHOR

James Embree began flying at the age of 22 while serving in the U.S. Army. After release from active duty in 1967 he pursued an aviation career and quickly earned Commercial, Mulit-Engine, Instrument and Instructor ratings.

This accomplished, he secured a position flight instructing at a pilot training school, where he soon became Chief Flight and Ground Instructor.

After spending four years in this position, Embree accepted a position as Chief Pilot for a corporation operating aircraft ranging from a Baron to a Lear Jet. He has remained in this position for 12 years. He holds Airline Transport Pilot, Commercial, and Flight Instructor Ratings. He holds Type Ratings in Cessna Citaitons and Lear Jets, and is an F.A.A. Designated Proficiency Pilot Examiner for Cessna Citation Aircraft.

TABLE OF CONTENTS

INTRODUCTION

During the years I have spent as a professional pilot, I have been exposed to many ways of doing the same thing. Some were worthless and others were great little bits of knowledge that I use every time I fly.

A few years ago, I bought a pocket computer and began to program it to accomplish some of those bothersome and time consuming jobs that must be done in the planning and execution of any flight. As time went on, my programs become more involved and I knew I had only scratched the surface. I had also run out of sources of information! So, I began a search for information, and found bits here and there, from friends, from the F.A.A., Air Force and Navy Manuals and from Government publications. This information was scattered everywhere and none of it very easy to find.

The purpose of this book is to save you the agony of trying to find all this information. I have lumped together in this book as much useful information as possible.

You will find Rules Of Thumb, Formulas, Graphs, Tables, and in many cases, Basic Language Programs for each of the subjects discussed.

If you own a "pocket" or "briefcase" computer or programable calculator, you may want to use the information in this book to write your own programs.

Don't let this talk of computers scare you off! This book is designed to be used alone and is a good cockpit friend.

One word of caution! This book is not a great literary work. I am not a writer. I am just a pilot who wants to share this information with other pilots.

ADVISORY

The information contained herein is advisory in nature and it is the responsibility of the user to determine that procedures used are not in conflict with the Airplane Flight Manual, Federal Aviation Regulations, Air Traffic Control Clearances or Safe Operating Practice.

The user is further advised that this information is general in nature and neither the Author nor the Publisher make any claim as to the suitability of this material to any particular Make or Model of Aircraft.

CHAPTER 1

TAKEOFF PERFORMANCE

CHAPTER CONTENTS

TAKEOFF PERFORMANCE

In this chapter we will explore the effects of the takeoff variables Wind, Density Altitude and Aircraft Weight on takeoff performance. If you are like most of us, you simply estimate these effects and go. Most of the time this will work, but I'm sure you've heard stories of the guy at the airport across town who bent his bird trying to takeoff when the combined effects of these takeoff variables were just too much for the airplane to handle. We have to believe that this event was not in the flight plan, so we ask ourselves the question "why?". Part of the answer, I believe, is that it is difficult to determine the actual conditions existing at the time of takeoff. For instance, the airport elevation is known and so is the temprature; but what is the density altitude? Or, the wind direction and velocity are known; but what portion of the wind is headwind and what portion is 90 degree crosswind?

This chapter will show you how to make quick and accurate assements of takeoff variables and determine their effects on takeoff distance and airplane control.

All you have to do is work from the start of the chapter to the end. Use the rules of thumb, tables or graphs as needed. Remember, it is not necessary to use each form of infromation supplied; only the one you like best.

CROSSWIND AND HEADWIND COMPONENTS

The force of a crosswind acting on an airplane is actually divided into two forces (components); one is the 90 degree crosswind force and the other is the direct headwind, or tailwind, force. The headwind or tailwind force is equal to the Cosine of the wind angle times the wind velocity.

The 90 degree crosswind force is equal to the Sine of the wind angle times the wind velocity. Use the table in Figure 1-1 or the graph in Figure 1-2 to determine the the values of these forces.

CROSSWIND AND HEADWIND MULTIPLIERS

To determine crosswind and headwind force, obtain the multipliers from the table in figure 1-1 then multiply them by the wind velocity (this information also works for enroute winds, more on this in a later chapter).

EXAMPLE:

Wind Angle = 30 degrees

Wind Speed = 20 knots

Crosswind Multiplier = .50 (from table)

Crosswind Force = .50 x 20

Crosswind Force = 10 knots acting as a 90 degree crosswind

Headwind Multiplier = .85 (from table).

Headwind Force = .85 x 20

Headwind Force = 17 knots acting as a headwind (or tailwind)

WIND COMPONENT MULTIPLIER TABLE		
WIND ANGLE	CROSSWIND Sine	HEADWIND Cosine
0	0.0	1.0
10	.20	1.0
20	.35	.90
30	.50	.85
40	.65	.75
50	.75	.65
60	.85	.50
70	.90	.35
80	1.0	.20
90	1.0	0.0

Figure 1-1

CROSSWIND COMPONENT GRAPH

Use the Graph in Figure 1-2 to determine the crosswind and headwind force. In the example, simply move along the diagonal line representing the wind angle (40 degrees) until it intercepts the circular line representing the wind velocity (point 3). Then read the tailwind or headwind force on the left of the chart (point 4) and the crosswind force at the bottom of the Graph (point 5).

CROSSWIND/HEADWIND FORCE BASIC LANGUAGE PROGRAM

```
10 INPUT"WIND ANGLE=";A,"WIND SPEED=";S
20 H=INT((100*(COS A*S))+.5)/100:
   X=INT((100*(SIN A*S))+.5)/100
30 PRINT"XWIND=";X;"  HDWND=";H:END
```

CROSSWIND COMPONENT GRAPH

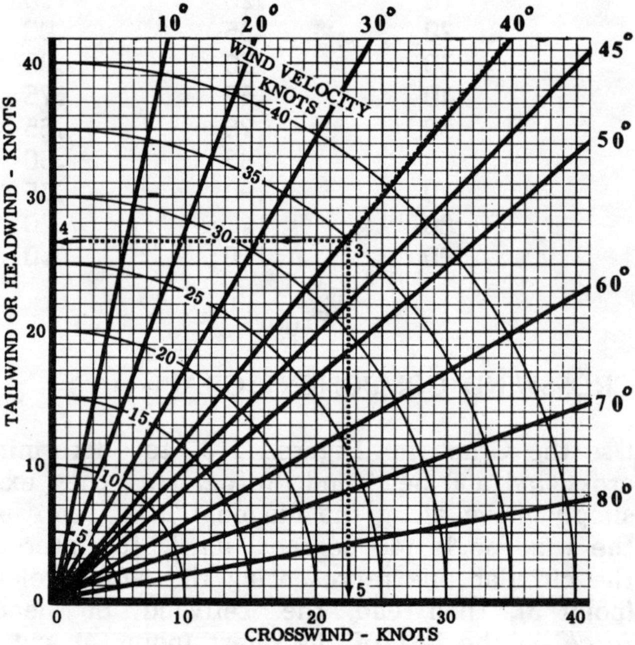

Figure 1-2

EXAMPLE: Wind angle 40 degrees. Wind speed 35 kts. Crosswind = 22.5 kts. Headwind or Tailwind = 26.8 kts.

DENSITY ALTITUDE CALCULATIONS

To determine Density Altitude you need Pressure Altitude and Temperature.

The easiest way to determine pressure altitude is to set the altimeter to 29.92 and read the value displayed by the hands of the instrument. The easiest way to determine the temperature is to listen to ATIS or get it from the Tower or Flight Service Station.

Since not all take off planning happens at the airport, and most of us don't carry altimeters around in our pockets, I offer the following information to aid you with your take off planning. I have supplied a temperature conversion graph here as well as other places in the book. This is to reduce the amount of page turning you have to do. See Figure 1-4.

Rule Of Thumb #1 — Atmospheric Pressure decreases at the rate of 1" Hg (1" Mercury) per 1000 feet of altitude above sea level. Therefore, to determine the pressure altitude, find the difference between 29.92 (Standard Sea Level Pressure) and the existing pressure; then multiply this value by 1000. This is the number of feet the pressure altitude varies from the field elevation. Now, if the existing pressure is higher than 29.92, subtract the altitude variation from the field elevation. If the existing pressure is lower than 29.92 then add the altitude variation to the field elevation.

EXAMPLE:

Field Elevation = 1230 feet

Existing Pressure = 30.16

Then:

30.16 - 29.92 = .24

.24 x 1000 = 240

30.16 is higher than 29.92

So:

Pressure Altitude = 1230 - 240

Pressure Altitude = 990 feet

The Table in Figure 1-3 gives the value to add or subtract to convert Field Elevation to Pressure Altitude. You will notice that one inch Mercury is not in reality equal to 1000 feet, but for most calculations one inch Mercury equals 1000 feet is quite satisfactory.

If you are like me you will use the table because it saves some mental gymnastics. Round off to the nearest .10 inches. This rounding will cause a maximum error of 50 feet.

TABLE FOR PRESSURE ALTITUDE

APPLY ALTITUDE CORRECTION TO FIELD ELEVATION			
STATION PRESSURE	ALTITUDE CORRECTION	STATION PRESSURE	ALTITUDE CORRECTION
28.00	+1,825	29.60	+300
28.10	+1,725	29.70	+205
28.20	+1,630	29.80	+110
28.30	+1,535	29.90	+20
28.40	+1,435	29.92	0
28.50	+1,340	30.00	-75
28.60	+1,245	30.10	-165
28.70	+1,150	30.20	-225
28.80	+1,050	30.30	-350
28.90	+955	30.40	-440
29.00	+865	30.50	-530
29.10	+770	30.60	-620
29.20	+675	30.70	-710
29.30	+580	30.80	-805
29.40	+485	30.90	-895
29.50	+390	31.00	-965

Figure 1-3

TEMPERATURE CONVERSION GRAPH

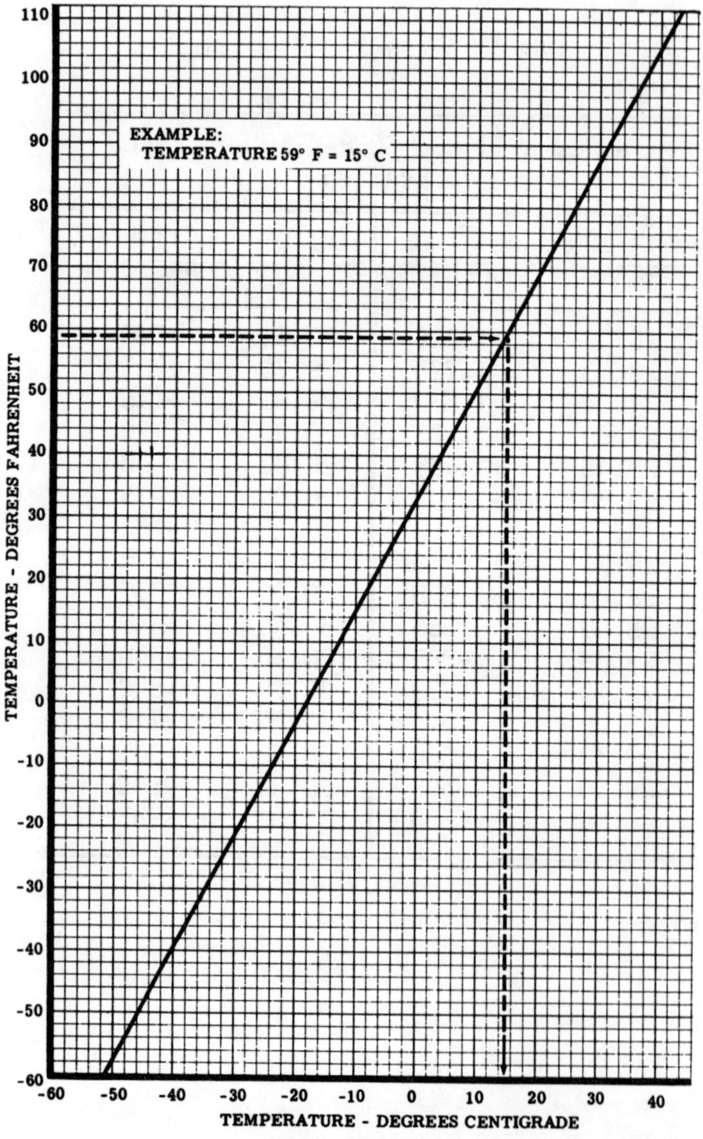

Figure 1-4

Rule Of Thumb #2 — To determine the Standard Celsius Temperature for Pressure Altitude; double the altitude in thousands of feet and subtract 15 then change the sign.

EXAMPLE:

Pressure Altitude = 1000

Double Altitude in Thousands = 2

Subtract = 2 - 15 = -13

Change Sign -13 = +13

Standard Temperature C = +13

STANDARD TEMPERATURE (C)
BASIC LANGUAGE PROGRAM

```
10 INPUT"PRESS. ALT =";A:D=15-(A*
   .001981)
20 PRINT"STD. TEMP.=";D;" C":END
```

Rule Of Thumb #3 — To determine the Standard Fahrenheit Temperature for Pressure Altitude; multiply the altitude in thousands by 3.5 and subtract this value from 59.

EXAMPLE:

Pressure Altitude = 2000

Altitude in Thousands = 2

Multiply = 2 X 3.5 = 7

Subtract = 59 - 7 = 52

Standard Temperature F = 52

STANDAND TEMPERATURE (F)
BASIC LANGUAGE PROGRAM

```
10 INPUT"PRESSURE ALTITUDE =";A:D=5
   9-(.003566*A)
20 PRINT"STANDARD TEMPERATURE =";D:
   END
```

See the graph in Figure 1-5 for Standard Temperature at Altitude.

Rule Of Thumb #4 -- The Density Altitude increases or decreases 630 feet for each 10 degrees Fahrenheit the temperature varies from Standard Temperature for a particular altitude.

NOTE: Use Pressure Altitude and Temperature information from preceding pages to determine values needed for actual calculations.

EXAMPLE:

Pressure Altitude = 1000

Standard Temperature = 55.4 F

Actual Temperature = 70 F

Temperature Variation = 14.6 F

$$\text{Altitude Variation} = \frac{14.6}{10} = 1.46$$

$$\begin{aligned}\text{Density Altitude} &= 630 \times 1.46 \\ &= 919.80\end{aligned}$$

$$\begin{aligned}\text{Density Altitude} &= 919.8 + 1000 \\ &= 1919.8\end{aligned}$$

The altitude variation is actually somewhat more than 630 feet per 10 degrees Fahrenheit at the lower pressure altitudes and somewhat less than 630 feet per 10 degrees Fahrenheit at the higher pressure altitudes, the average turns out to be about 630 feet per 10 degrees Fahrenheit.

Use graph in Figure 1-6 to determine the Density Altitude.

The tables in Figure 1-7 provide Density Altitudes for Pressure Altitudes from sea level to 10,000 feet at 500 foot intervals and 5 degree Fahrenheit temperature intervals. No Density Altitudes below sea level are provided. If Density Altitude is below sea level, use sea level performance data.

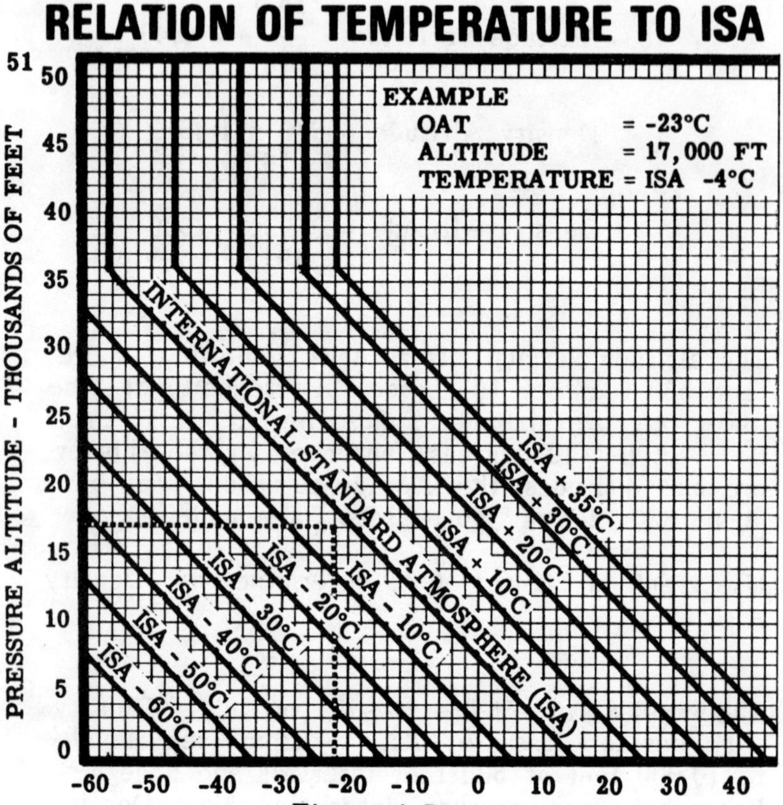

Figure 1-5

To use the Graph in Figure 1-6 just enter the Graph at the bottom with either the Fahrenheit or Celsius temperature and move upward to the diagonal Pressure Altitude line then directly left and read the Density altitude.

In the example 75 degrees F and 4000 feet Pressure Altitude equals 5923 feet Density Altitude. No one I know can read this kind of graph with this accuracy....if you must know, I cheated and used the tables in Figure 1-7 to work out this example. The values obtained from this graph will be accurate enough for takeoff calculations.

DENSITY ALTITUDE GRAPH

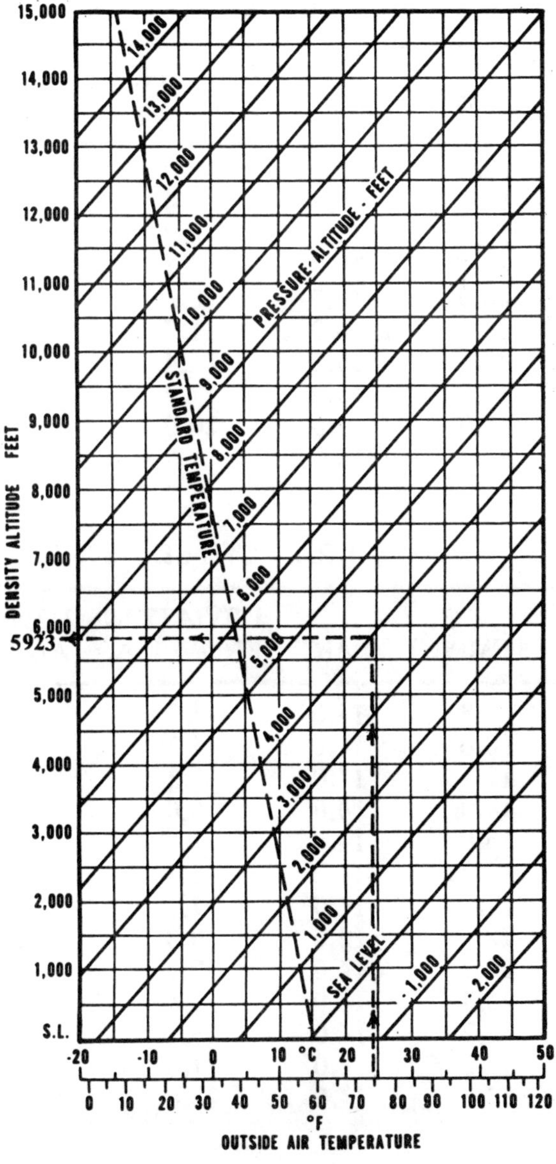

Figure 1-6

PRESSURE ALTITUDE 0

TEMP(F)	TEMP(C)	ISA(C)	DEVIATION FROM ISA	DENSITY ALTITUDE
59.0	15.0	15.0	0.0	0
60.0	15.6	15.0	0.6	66
65.0	18.3	15.0	3.3	393
70.0	21.1	15.0	6.1	715
75.0	23.9	15.0	8.9	1035
80.0	26.7	15.0	11.7	1350
85.0	29.4	15.0	14.4	1662
90.0	32.2	15.0	17.2	1970
95.0	35.0	15.0	20.0	2275
100.0	37.8	15.0	22.8	2576

PRESSURE ALTITUDE 500

TEMP(F)	TEMP(C)	ISA(C)	DEVIATION FROM ISA	DENSITY ALTITUDE
49.7	9.8	14.0	-4.2	0
50.0	10.0	14.0	-4.0	20
55.0	12.8	14.0	-1.2	354
58.2	14.0	14.0	0.0	500
60.0	15.6	14.0	1.6	682
65.0	18.3	14.0	4.3	1008
70.0	21.1	14.0	7.1	1329
75.0	23.9	14.0	9.9	1647
80.0	26.7	14.0	12.7	1962
85.0	29.4	14.0	15.4	2272
90.0	32.2	14.0	18.2	2579
95.0	35.0	14.0	21.0	2882
100.0	37.8	14.0	23.8	3183

Figure 1-7

PRESSURE ALTITUDE 1000

TEMP(F)	TEMP(C)	ISA(C)	DEVIATION FROM ISA	DENSITY ALTITUDE
40.5	4.7	13.0	-8.3	0
45.0	7.2	13.0	-5.8	304
50.0	10.0	13.0	-3.0	640
55.0	12.8	13.0	-0.2	971
55.4	13.0	13.0	0.0	1000
60.0	15.6	13.0	2.6	1299
65.0	18.3	13.0	5.3	1623
70.0	21.1	13.0	8.1	1943
75.0	23.9	13.0	10.9	2260
80.0	26.7	13.0	13.7	2572
85.0	29.4	13.0	16.4	2882
90.0	32.2	13.0	19.2	3188
95.0	35.0	13.0	22.0	3490
100.0	37.8	13.0	24.8	3789

PRESSURE ALTITUDE 1500

TEMP(F)	TEMP(C)	ISA(C)	DEVIATION FROM ISA	DENSITY ALTITUDE
31.5	-0.3	12.0	-12.3	0
35.0	1.7	12.0	-10.3	243
40.0	4.4	12.0	-7.6	585
45.0	7.2	12.0	-4.8	924
50.0	10.0	12.0	-2.0	1258
53.7	12.0	12.0	0.0	1500
55.0	12.8	12.0	0.8	1589
60.0	15.6	12.0	3.6	1915
65.0	18.3	12.0	6.3	2238
70.0	21.1	12.0	9.1	2557
75.0	23.9	12.0	11.9	2872

Figure 1-7

PRESSURE ALTITUDE 1500

TEMP(F)	TEMP(C)	ISA(C)	DEVIATION FROM ISA	DENSITY ALTITUDE
80.0	26.7	12.0	14.7	3183
85.0	29.4	12.0	17.4	3491
90.0	32.2	12.0	20.2	3796
95.0	35.0	12.0	23.0	4097
100.0	37.8	12.0	25.8	4394

PRESSURE ALTITUDE 2000

TEMP(F)	TEMP(C)	ISA(C)	DEVIATION FROM ISA	DENSITY ALTITUDE
22.6	-5.2	11.0	-16.2	0
25.0	-3.9	11.0	-14.9	170
30.0	-1.1	11.0	-12.1	520
35.0	1.7	11.0	-9.3	865
40.0	4.4	11.0	-6.6	1207
45.0	7.2	11.0	-3.8	1544
50.0	10.0	11.0	-1.0	1877
51.9	11.0	11.0	0.0	2000
55.0	12.8	11.0	1.8	2206
60.0	15.6	11.0	4.6	2531
65.0	18.3	11.0	7.3	2852
70.0	21.1	11.0	10.0	3169
75.0	23.9	11.0	12.9	3483
80.0	26.7	11.0	15.7	3793
85.0	29.4	11.0	18.4	4100
90.0	32.2	11.0	21.2	4403
95.0	35.0	11.0	24.0	4703
100.0	37.8	11.0	26.7	4999

Figure 1-7

PRESSURE ALTITUDE 2500

TEMP(F)	TEMP(C)	ISA(C)	DEVIATION FROM ISA	DENSITY ALTITUDE
13.8	-10.1	10.0	-20.1	0
15.0	-9.5	10.0	-19.5	85
20.0	-6.7	10.0	-16.7	442
25.0	-3.9	10.0	-13.9	795
30.0	-1.1	10.0	-11.1	1143
35.0	1.7	10.0	-8.3	1488
40.0	4.4	10.0	-5.6	1827
45.0	7.2	10.0	-2.8	2163
50.0	10.0	10.0	0.0	2500
55.0	12.8	10.0	2.8	2822
60.0	15.6	10.0	5.6	3146
65.0	18.3	10.0	8.3	3465
70.0	21.1	10.0	11.1	3781
75.0	23.9	10.0	13.9	4094
80.0	26.7	10.0	16.7	4403
85.0	29.4	10.0	19.4	4708
90.0	32.2	10.0	22.2	5010
95.0	35.0	10.0	25.0	5308
100.0	37.8	10.0	27.8	5603

PRESSURE ALTITUDE 3000

TEMP(F)	TEMP(C)	ISA(C)	DEVIATION FROM ISA	DENSITY ALTITUDE
5.2	-14.9	9.1	-24.0	0
10.0	-12.2	9.1	-21.3	352
15.0	-9.5	9.1	-18.6	712
20.0	-6.7	9.1	-15.8	1608
25.0	-3.9	9.1	-13.0	1420
30.0	-1.1	9.1	-10.2	1767
35.0	1.7	9.1	-7.4	2109

Figure 1-7

PRESSURE ALTITUDE 3000

TEMP(F)	TEMP(C)	ISA(C)	DEVIATION FROM ISA	DENSITY ALTITUDE
40.0	4.4	9.1	-4.7	2447
45.0	7.2	9.1	-1.9	2782
48.3	9.1	9.1	0.0	3000
50.0	10.0	9.1	0.9	3112
55.0	12.8	9.1	3.7	3438
60.0	15.6	9.1	6.5	3760
65.0	18.3	9.1	9.2	4078
70.0	21.1	9.1	12.0	4393
75.0	23.9	9.1	14.8	4704
80.0	26.7	9.1	17.6	5012
85.0	29.4	9.1	20.3	5316
90.0	32.2	9.1	23.1	5616
95.0	35.0	9.1	25.9	5913
100.0	37.8	9.1	28.7	6207

PRESSURE ALTITUDE 3500

TEMP(F)	TEMP(C)	ISA(C)	DEVIATION FROM ISA	DENSITY ALTITUDE
0.0	-17.8	8.1	-25.9	248
5.0	-15.0	8.1	-23.1	617
10.0	-12.2	8.1	-20.3	980
15.0	-9.5	8.1	-17.6	1339
20.0	-6.7	8.1	-14.8	1694
25.0	-3.9	8.1	-12.0	2044
30.0	-1.1	8.1	-9.2	2389
35.0	1.7	8.1	-6.4	2730
40.0	4.4	8.1	-3.7	3067
45.0	7.2	8.1	-0.9	3400

Figure 1-7

PRESSURE ALTITUDE 3500

TEMP(F)	TEMP(C)	ISA(C)	DEVIATION FROM ISA	DENSITY ALTITUDE
46.5	8.1	8.1	0.0	3500
50.0	10.0	8.1	1.9	3728
55.0	12.8	8.1	4.7	4053
60.0	15.6	8.1	7.5	4374
65.0	18.3	8.1	10.2	4691
70.0	21.1	8.1	13.0	5004
75.0	23.9	8.1	15.8	5314
80.0	26.7	8.1	18.6	5620
85.0	29.4	8.1	21.3	5923
90.0	32.2	8.1	24.1	6222
95.0	35.0	8.1	26.9	6518
100.0	37.8	8.1	29.7	6811

PRESSURE ALTITUDE 4000

TEMP(F)	TEMP(C)	ISA(C)	DEVIATION FROM ISA	DENSITY ALTITUDE
0.0	-17.8	7.1	-24.9	879
5.0	-15.0	7.1	-22.1	1246
10.0	-12.2	7.1	-19.3	1609
15.0	-9.5	7.1	-16.6	1966
20.0	-6.7	7.1	-13.8	2319
25.0	-3.9	7.1	-11.0	2667
30.0	-1.1	7.1	-8.2	3011
35.0	1.7	7.1	-5.4	3351
40.0	4.4	7.1	-2.7	3686
44.7	7.1	7.1	0.0	4000
45.0	7.2	7.1	0.1	4017
50.0	10.0	7.1	2.9	4345
55.0	12.8	7.1	5.7	4668
60.0	15.6	7.1	8.5	4987

Figure 1-7

PRESSURE ALTITUDE 4000

TEMP(F)	TEMP(C)	ISA(C)	DEVIATION FROM ISA	DENSITY ALTITUDE
65.0	18.3	7.1	11.2	5303
70.0	21.1	7.1	14.0	5615
75.0	23.9	7.1	16.8	5923
80.0	26.7	7.1	19.6	6228
85.0	29.4	7.1	22.3	6529
90.0	32.2	7.1	25.1	6827
95.0	35.0	7.1	27.9	7122
100.0	37.8	7.1	30.7	7413

PRESSURE ALTITUDE 4500

TEMP(F)	TEMP(C)	ISA(C)	DEVIATION FROM ISA	DENSITY ALTITUDE
0.0	-17.8	6.1	-23.9	1510
5.0	-15.0	6.1	-21.1	1876
10.0	-12.2	6.1	-18.3	2236
15.0	-9.5	6.1	-15.6	2592
20.0	-6.7	6.1	-12.8	2943
25.0	-3.9	6.1	-10.0	3290
30.0	-1.1	6.1	-7.2	3632
35.0	1.7	6.1	-4.4	3971
40.0	4.4	6.1	-1.7	4304
42.9	6.1	6.1	0.0	4500
45.0	7.2	6.1	1.1	4634
50.0	10.0	6.1	3.9	4960
55.0	12.8	6.1	6.7	5282
60.0	15.6	6.1	9.5	5600
65.0	18.3	6.1	12.2	5915
70.0	21.1	6.1	15.0	6225
75.0	23.9	6.1	17.8	6532
80.0	26.7	6.1	20.6	6836
85.0	29.4	6.1	23.3	7136
90.0	32.2	6.1	26.1	7432
95.0	35.0	6.1	28.9	7726
100.0	37,8	6.1	31.7	8016

Figure 1-7

PRESSURE ALTITUDE 5000

TEMP(F)	TEMP(C)	ISA(C)	DEVIATION FROM ISA	DENSITY ALTITUDE
0.0	-17.8	5.1	-22.9	2141
5.0	-15.0	5.1	-20.1	2504
10.0	-12.2	5.1	-17.3	2863
15.0	-9.5	5.1	-14.6	3218
20.0	-6.7	5.1	-11.8	3568
25.0	-3.9	5.1	-9.0	3913
30.0	-1.1	5.1	-6.2	4254
35.0	1.7	5.1	-3.4	4590
40.0	4.4	5.1	-0.7	4923
41.1	5.1	5.1	0.0	5000
45.0	7.2	5.1	2.1	5251
50.0	10.0	5.1	4.9	5576
55.0	12.8	5.1	7.7	5896
60.0	15.6	5.1	10.5	6213
65.0	18.3	5.1	13.2	6526
70.0	21.1	5.1	16.0	6835
75.0	23.9	5.1	18.8	7141
80.0	26.7	5.1	21.6	7443
85.0	29.4	5.1	24.3	7741
90.0	32.2	5.1	27.1	8037
95.0	35.0	5.1	29.9	8321
100.0	37.8	5.1	32.7	8618

PRESSURE ALTITUDE 5500

TEMP(F)	TEMP(C)	ISA(C)	DEVIATION FROM ISA	DENSITY ALTITUDE
0.0	-17.8	4.1	-21.9	2770
5.0	-15.0	4.1	-19.1	3133
10.0	-12.2	4.1	-16.3	3490
15.0	-9.5	4.1	-13.6	3843
20.0	-6.7	4.1	-10.8	4194

Figure 1-7

PRESSURE ALTITUDE 5500

TEMP(F)	TEMP(C)	ISA(C)	DEVIATION FROM ISA	DENSITY ALTITUDE
25.0	-3.9	4.1	-8.0	4535
30.0	-1.1	4.1	-5.2	4874
35.0	1.7	4.1	-2.4	5309
39.3	4.1	4.1	0.0	5500
40.0	4.4	4.1	0.3	5540
45.0	7.2	4.1	3.1	5867
50.0	10.0	4.1	5.9	6190
55.0	12.8	4.1	8.7	6509
60.0	15.6	4.1	11.5	6835
65.0	18.3	4.1	14.2	7136
70.0	21.1	4.1	17.0	7444
75.0	23.9	4.1	19.8	7748
80.0	26.7	4.1	22.6	8049
85.0	29.4	4.1	25.3	8347
90.0	32.2	4.1	28.1	8641
95.0	35.0	4.1	30.9	8931
100.0	37.8	4.1	33.7	9219

PRESSURE ALTITUDE 6000

TEMP(F)	TEMP(C)	ISA(C)	DEVIATION FROM ISA	DENSITY ALTITUDE
0.0	-17.8	3.1	-20.9	4000
5.0	-15.0	3.1	-18.1	3760
10.0	-12.2	3.1	-15.3	4116
15.0	-9.5	3.1	-12.6	4476
20.0	-6.7	3.1	-9.8	4814
25.0	-3.9	3.1	-7.0	5156
30.0	-1.1	3.1	-4.2	5494
35.0	1.7	3.1	-1.4	5828
37.5	3.1	3.1	0.0	6000
40.0	4.4	3.1	1.3	6157

Figure 1-7

PRESSURE ALTITUDE 6000

TEMP(F)	TEMP(C)	ISA(C)	DEVIATION FROM ISA	DENSITY ALTITUDE
45.0	7.2	3.1	4.1	6483
50.0	10.0	3.1	6.9	6804
55.0	12.8	3.1	9.7	7122
60.0	15.6	3.1	12.5	7436
65.0	18.3	3.1	15.2	7746
70.0	21.1	3.1	18.0	8053
75.0	23.9	3.1	20.8	8356
80.0	26.7	3.1	23.6	8655
85.0	29.4	3.1	26.3	8951
90.0	32.2	3.1	29.1	9244
95.0	35.0	3.1	31.9	9533
100.0	37.8	3.1	34.7	9820

PRESSURE ALTITUDE 6500

TEMP(F)	TEMP(C)	ISA(C)	DEVIATION FROM ISA	DENSITY ALTITUDE
0.0	-17.8	2.1	-19.9	4028
5.0	-15.0	2.1	-17.1	4387
10.0	-12.2	2.1	-14.3	4741
15.0	-9.5	2.1	-11.6	5091
20.0	-6.7	2.1	-8.8	5437
25.0	-3.9	2.1	-6.0	5777
30.0	-1.1	2.1	-3.2	6114
35.0	1.7	2.1	-0.4	6446
35.7	2.1	2.1	0.0	6500
40.0	4.4	2.1	2.3	6774
45.0	7.2	2.1	5.1	7098
50.0	10.0	2.1	7.9	7418
55.0	12.8	2.1	10.7	7734
60.0	15.6	2.1	13.5	8047
65.0	18.3	2.1	16.2	8356

Figure 1-7

PRESSURE ALTITUDE 6500

TEMP(F)	TEMP(C)	ISA(C)	DEVIATION FROM ISA	DENSITY ALTITUDE
70.0	21.1	2.1	19.0	8661
75.0	23.9	2.1	21.8	8962
80.0	26.7	2.1	24.6	9261
85.0	29.4	2.1	27.3	9555
90.0	32.2	2.1	30.1	9846
95.0	35.0	2.1	32.9	10135
100.0	37.8	2.1	35.7	10420

PRESSURE ALTITUDE 7000

TEMP(F)	TEMP(C)	ISA(C)	DEVIATION FROM ISA	DENSITY ALTITUDE
0.0	-17.8	1.1	-18.9	4657
5.0	-15.0	1.1	-16.1	5014
10.0	-12.2	1.1	-13.3	5367
15.0	-9.5	1.1	-10.6	5715
20.0	-6.7	1.1	-7.8	6058
25.0	-3.9	1.1	-5.0	6398
30.0	-1.1	1.1	-2.2	6733
33.9	1.1	1.1	0.0	7000
35.0	1.7	1.1	0.6	7063
40.0	4.4	1.1	3.3	7390
45.0	7.2	1.1	6.1	7713
50.0	10.0	1.1	8.9	8031
55.0	12.8	1.1	11.7	8346
60.0	15.6	1.1	14.5	8657
65.0	18.3	1.1	17.2	8965
70.0	21.1	1.1	20.0	9268
75.0	23.9	1.1	22.8	9567
80.0	26.7	1.1	25.6	9866
85.0	29.4	1.1	28.3	10159
90.0	32.2	1.1	31.1	10449
95.0	35.0	1.1	33.9	10736
100.0	37.8	1.1	36.7	11020

Figure 1-7

PRESSURE ALTITUDE 7500

TEMP(F)	TEMP(C)	ISA(C)	DEVIATION FROM ISA	DENSITY ALTITUDE
0.0	-17.8	0.1	-17.9	5284
5.0	-15.0	0.1	-15.1	5640
10.0	-12.2	0.1	-12.3	5991
15.0	-9.5	0.1	-9.6	6338
20.0	-6.7	0.1	-6.8	6680
25.0	-3.9	0.1	-4.0	7018
30.0	-1.1	0.1	-1.2	7351
32.0	0.1	0.1	0.0	7500
35.0	1.7	0.1	1.6	7680
40.0	4.4	0.1	4.3	8005
45.0	7.2	0.1	7.1	8327
50.0	10.0	0.1	9.9	8644
55.0	12.8	0.1	12.7	8957
60.0	15.6	0.1	15.5	9267
65.0	18.3	0.1	18.2	9573
70.0	21.1	0.1	21.0	9875
75.0	23.9	0.1	23.8	10174
80.0	26.7	0.1	26.6	10470
85.0	29.4	0.1	29.3	10762
90.0	32.2	0.1	32.1	11051
95.0	35.0	0.1	34.9	11337
100.0	37.8	0.1	37.7	11619

PRESSURE ALTITUDE 8000

TEMP(F)	TEMP(C)	ISA(C)	DEVIATION FROM ISA	DENSITY ALTITUDE
0.0	-17.8	-0.8	-17.0	5911
5.0	-15.0	-0.8	-14.2	6266
10.0	-12.2	-0.8	-11.4	6615
15.0	-9.5	-0.8	-8.7	6960
20.0	-6.7	-0.8	-5.9	7301

Figure 1-7

PRESSURE ALTITUDE 8000

TEMP(F)	TEMP(C)	ISA(C)	DEVIATION FROM ISA	DENSITY ALTITUDE
25.0	-3.9	-0.8	-3.1	7637
30.0	-1.1	-0.8	-0.3	7969
30.5	-0.8	-0.8	0.0	8000
35.0	1.7	-0.8	2.5	8297
40.0	4.4	-0.8	5.2	8620
45.0	7.2	-0.8	8.0	8940
50.0	10.0	-0.8	10.8	9256
55.0	12.8	-0.8	13.6	9568
60.0	15.6	-0.8	16.4	9876
65.0	18.3	-0.8	19.1	10181
70.0	21.1	-0.8	21.9	10482
75.0	23.9	-0.8	24.7	10780
80.0	26.7	-0.8	27.5	11074
85.0	29.4	-0.8	30.2	11364
90.0	32.2	-0.8	33.0	11625
95.0	35.0	-0.8	35.8	11937
100.0	37.8	-0.8	38.6	12218

PRESSURE ALTITUDE 8500

TEMP(F)	TEMP(C)	ISA(C)	DEVIATION FROM ISA	DENSITY ALTITUDE
0.0	-17.8	-1.8	-16.0	6538
5.0	-15.0	-1.8	-13.2	6891
10.0	-12.2	-1.8	-10.4	7239
15.0	-9.5	-1.8	-7.7	7582
20.0	-6.7	-1.8	-4.9	7921
25.0	-3.9	-1.8	-2.1	8256
28.7	-1.8	-1.8	0.0	8500
30.0	-1.1	-1.8	0.7	8586
35.0	1.7	-1.8	3.5	8913
40.0	4.4	-1.8	6.2	9235

Figure 1-7

PRESSURE ALTITUDE 8500

TEMP(F)	TEMP(C)	ISA(C)	DEVIATION FROM ISA	DENSITY ALTITUDE
45.0	7.2	-1.8	9.0	9553
50.0	10.0	-1.8	11.8	9868
55.0	12.8	-1.8	14.6	10178
60.0	15.6	-1.8	17.4	10485
65.0	18.3	-1.8	20.1	10788
70.0	21.1	-1.8	22.9	11088
75.0	23.9	-1.8	25.7	11384
80.0	26.7	-1.8	28.5	11677
85.0	29.4	-1.8	31.2	11967
90.0	32.2	-1.8	34.0	12253
95.0	35.0	-1.8	36.8	12536
100.0	37.8	-1.8	39.6	12816

PRESSURE ALTITUDE 9000

TEMP(F)	TEMP(C)	ISA(C)	DEVIATION FROM ISA	DENSITY ALTITUDE
0.0	-17.8	-2.8	-15.0	7164
5.0	-15.0	-2.8	-12.2	7515
10.0	-12.2	-2.8	-9.4	7862
15.0	-9.5	-2.8	-6.7	8203
20.0	-6.7	-2.8	-3.9	8541
25.0	-3.9	-2.8	-1.1	8874
26.9	-2.8	-2.8	0.0	9000
30.0	-1.1	-2.8	1.7	9203
35.0	1.7	-2.8	4.5	9528
40.0	4.4	-2.8	7.2	9849
45.0	7.2	-2.8	10.0	10166
50.0	10.0	-2.8	12.8	10479
55.0	12.8	-2.8	15.6	10787
60.0	15.6	-2.8	18.4	11093

Figure 1-7

PRESSURE ALTITUDE 9000

TEMP(F)	TEMP(C)	ISA(C)	DEVIATION FROM ISA	DENSITY ALTITUDE
65.0	18.3	-2.8	21.1	11395
70.0	21.1	-2.8	23.9	11694
75.0	23.9	-2.8	26.7	11989
80.0	26.7	-2.8	29.5	12280
85.0	29.4	-2.8	32.2	12568
90.0	32.2	-2.8	35.0	12853
95.0	35.0	-2.8	37.8	13135
100.0	37.8	-2.8	40.6	13414

PRESSURE ALTITUDE 9500

TEMP(F)	TEMP(C)	ISA(C)	DEVIATION FROM ISA	DENSITY ALTITUDE
0.0	-17.8	-3.8	-14.0	7790
5.0	-15.0	-3.8	-11.2	8139
10.0	-12.2	-3.8	-8.4	8484
15.0	-9.5	-3.8	-5.7	8824
20.0	-6.7	-3.8	-2.9	9160
25.0	-3.9	-3.8	-0.1	9492
25.1	-3.8	-3.8	0.0	9500
30.0	-1.1	-3.8	2.7	9819
35.0	1.7	-3.8	5.5	10143
40.0	4.4	-3.8	8.2	10462
45.0	7.2	-3.8	11.0	10777
50.0	10.0	-3.8	13.8	11089
55.0	12.8	-3.8	16.6	11397
60.0	15.6	-3.8	19.4	11701
65.0	18.3	-3.8	22.1	12002
70.0	21.1	-3.8	24.9	12299
75.0	23.9	-3.8	27.7	12592
80.0	26.7	-3.8	30.5	12883

Figure 1-7

PRESSURE ALTITUDE 9500

TEMP(F)	TEMP(C)	ISA(C)	DEVIATION FROM ISA	DENSITY ALTITUDE
85.0	29.4	-3.8	33.2	13170
90.0	32.2	-3.8	36.0	13453
95.0	35.0	-3.8	38.8	13734
100.0	37.8	-3.8	41.6	14011

PRESSURE ALTITUDE 10000

TEMP(F)	TEMP(C)	ISA(C)	DEVIATION FROM ISA	DENSITY ALTITUDE
0.0	-17.8	-4.8	-13.0	8415
5.0	-15.0	-4.8	-10.2	8762
10.0	-12.2	-4.8	-7.4	9106
15.0	-9.5	-4.8	-4.7	9445
20.0	-6.7	-4.8	-1.9	9779
23.3	-4.8	-4.8	0.0	10000
25.0	-3.9	-4.8	0.9	10109
30.0	-1.1	-4.8	3.7	10435
35.0	1.7	-4.8	6.5	10757
40.0	4.4	-4.8	9.2	11075
45.0	7.2	-4.8	12.0	11389
50.0	10.0	-4.8	14.8	11699
55.0	12.8	-4.8	17.6	12006
60.0	15.6	-4.8	20.4	12308
65.0	18.3	-4.8	23.1	12608
70.0	21.1	-4.8	25.9	12903
75.0	23.9	-4.8	28.7	13195
80.0	26.7	-4.8	31.5	13484
85.0	29.4	-4.8	34.2	13770
90.0	32.2	-4.8	37.0	14052
95.0	35.0	-4.8	39.8	14332
100.0	37.8	-4.8	42.6	14608

Figure 1-7

DENSITY ALTITUDE FORMULA

$$D=(145426 \times (1-(((288.16-P \times .001981) \div 288.16)^{5.2563} \div ((273.16+C) \div 288.16))^{.235}))$$

WHERE:

D = Density Altitude

P = Pressure Altitude

C = Degrees Celsius

DENSITY ALTITUDE BASIC LANGUAGE PROGRAM

```
10 INPUT"PRESS. ALT.=";P,"TEMP
   (F)=";F:C=(5/9)*(F-32)
20 DD=1-(((((288.16-(P*.001981))
   /288.16)↑5.2563)
   /((273.16+C)/288.16))↑.235)
30 D=145426*DD
35 TA=15-(P*.001981)
40 PRINT"TEMP. C=";C:PRINT"STD.
   TEMP=";TA
45 PRINT"DEN. ALT.=";D:END
```

This program provides the Celsius temperature for the Fahrenheit temperature input, the standard temperature for the Pressure Altitude input, as well as the Density Altitude.

EFFECT OF TAKEOFF VARIABLES

Once you have determined the Crosswind and Headwind, Density Altitude and Actual Takeoff Weight of the airplane, you may use the following section to determine the effect each of these variables will have on aircraft performance.
To use the information in this section, look up the Gross Weight, Sea Level Standard Day, Zero Wind takeoff distance for your airplane. Write this information down and keep it handy (there are several forms in the back of this book for this purpose). Now, use the Rules Of Thumb, Graphs, Tables or Formulas to determine the the approximate effects of existing conditions on the Gross Weight, Sea Level Standard Day, Zero Wind takeoff distance. Remember, use the method you like the best. The idea here is to find and use the method that you are most comfortable with and that saves you the most time.

Here are two Rules Of Thumb that can be quite handy if for some reason the Aircraft Operating Manual is not available.

Rule Of Thumb #5 — Lift Off Speed (Vlof) is about 1.15 times the Power Off Stall Speed (Vs1) or the Minimum Control Speed in the takeoff configuration.

EXAMPLE:

Power Off Stall Speed (clean) = 60 knots.

Vlof = 60 x 1.15 = 69

Vlof = 69 knots

Rule Of Thumb #6 -- The maximum 90 degree crosswind component of an airplane may be determined by multiplying Vso by 0.2.

EXAMPLE:

Vso = 55 knots

Crosswind Component = 55 x 0.2 = 11

Crosswind Component = 11 knots

This value is the minimum required by F.A.A. certification standards. Most airplanes have greater capability than indicated by this rule. Therefore, if the actual component is unknown this rule should give you a safety margin.

WEIGHT EFFECT

Note: The information presented is based on aircraft with power loading of 10 to 15 pounds per horsepower. The information should be properly biased to account for actual power loading.

Rule Of Thumb #7 -- The change in takeoff distance varies at least as the square of the change in Gross Weight.

Note: To determine the percent weight change divide the new weight by the old weight. To determine the percent of gross weight the actual weight represents divide actual weight by maximum allowed gross weight.

EXAMPLE:

Old Weight = 2000 pounds

New Weight = 2500 pounds

Takeoff distance at 2000 pounds = 1000 feet.

Weight change factor $= \frac{2500}{2000} = 1.25$

Squaring 1.25 x 1.25 = 1.56

New takeoff distance = 1.56 x 1000 = 1560 feet.

OR

Actual Weight = 2300 pounds

Maximum Weight Allowed = 2700 pounds

Takeoff distance at 2700 pounds = 1500 feet.

Weight change factor $= \frac{2300}{2700} = .85$

Squaring .85 x .85 = .72

New takeoff distance = .72 x 1500 = 1080 feet.

Rule Of Thumb #8 — A 10% change in aircraft gross weight will cause a 21% change in takeoff distance. Do not use this rule if weight change is excess of 50%.

EXAMPLE:

Weight Increase 10%

Takeoff distance before increase = 1000 feet.

New takeoff distance = 1000 + (1000 x .21) = 1210 feet.

New takeoff distance = 1210 feet.

WEIGHT Vs TAKEOFF DISTANCE
BASIC LANGUAGE PROGRAM

```
10 INPUT"OLD WEIGHT= ";O,"NEW WEIGHT= "
   ;N,"OLD T.O.DIST.=";D
20 A=((N/O)↑2)*D:PRINT"NEW DIST.=";A
30 END
```

USE GRAPH IN FIGURE 1-8 TO DETERMINE TAKEOFF DISTANCE CHANGE DUE TO WEIGHT CHANGE.

WEIGHT CHANGE Vs TAKEOFF DISTANCE

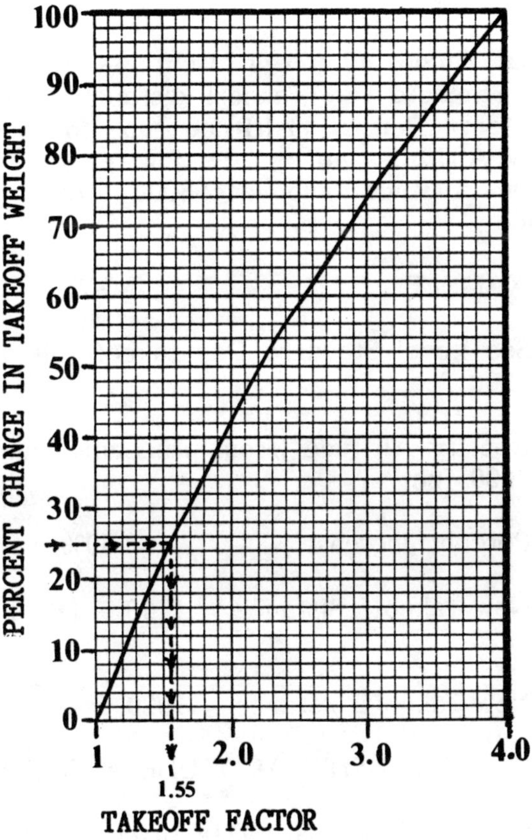

Example:
Weight change 25%. If weight change is an increase, multiply original takeoff distance by 1.55. If weight change is a **decrease** then **divide** the original takeoff distance by 1.55

Figure 1-8

DENSITY ALTITUDE EFFECT

Apply the following rules to the data derived from the peceding information or from the Flight Manual.

Rule Of Thumb #9 (fixed pitch) — Increase Sea Level, Standard Day Takeoff distance 15% for each 1000 feet Density Altitude up to 8000 feet Density Altitude and by 20% for each 1000 feet Density Altitude above 8000 feet Density Altitude.

EXAMPLE:

Sea Level Standard Day takeoff distance = 1200 feet

Density Altitude (from preceding charts) = 4000 feet

Percent Increase = 4 x 15 = 60%

New takeoff distance = 1200 + (1200 x .6) = 1920

New takeoff distance = 1920 feet

IF DENSITY ALTITUDE ABOVE 8000

Sea Level Standard Day takeoff distance = 1200 feet

Density Altitude (from preceding charts) = 9000 feet

Percent Increase to 8000 feet = 8 x 15 = 120%

Takeoff distance at 8000 feet = 1200 + (1200 x 1.20) = 2640

Percent Increase above 8000 feet = 1 x 20 = 20%

New takeoff distance = 2640 + (2640 x .20) = 3168

New takeoff distance = 3168 feet

Rule Of Thumb #10 (variable pitch) -- Increase Sea Level, Standard Day Takeoff distance 12% for each 1000 feet Density Altitude up to 6000 feet Density Altitude and 14% for each 1000 feet Density Altitude above 6000 feet Density Altitude.

EXAMPLE:

Sea Level, Standard Day takeoff distance = 1200 feet.

Density Altitude (from preceding charts) = 4000 feet.

Percent Increase = 4 x 12 = 48%

New takeoff distance = 1200 + (1200 x .48) = 1776

New takeoff distance = 1776 feet.

IF DENSITY ALTITUDE ABOVE 6000

Sea Level, Standard Day takeoff distance = 1200 feet.

Density Altitude (from preceding charts) = 9000 feet.

Percent Increase to 6000 feet = 6 x 12 = 72%

Takeoff distance at 6000 feet = 1200 + (1200 x .72) = 2064 feet.

Percent Increase above 6000 feet = 3 x 14 = 42%

New takeoff distance = 2064 + (2064 x .42) = 2931

New takeoff distance = 2931 feet

DEN. ALT. Vs T.O. DISTANCE
BASIC LANGUAGE PROGRAM

```
10 INPUT"FIXED(F) VARIABLE(V)";A$
20 INPUT"S.L. TO. DIST.= ";D,"DEN ALT.= ";A
30 IF A$<>"F"AND A$<>"V"THEN 10
40 IF A$="F"THEN LET X=.000155:
   Y=.000402:GOTO100
50 X=.000123:Y=.000188
100 ND=(1+(A*X))*D
110 IF A$="F"AND A>8000THEN LET
    ND=(2.24+((A-8000)*Y))*D
120 IF A$="V"AND A>5700THEN LET
    ND=(1.70+((A-5700)*Y))*D
200 PRINT "NEW DIST.= ";
    INT((ND*100)+.05)/100:END
```

YOU MAY ALSO USE THE GRAPH IN FIGURE 1-9 TO DETERMINE TAKEOFF MULTIPLIER.

DENSITY ALTITUDE Vs TAKEOFF DISTANCE

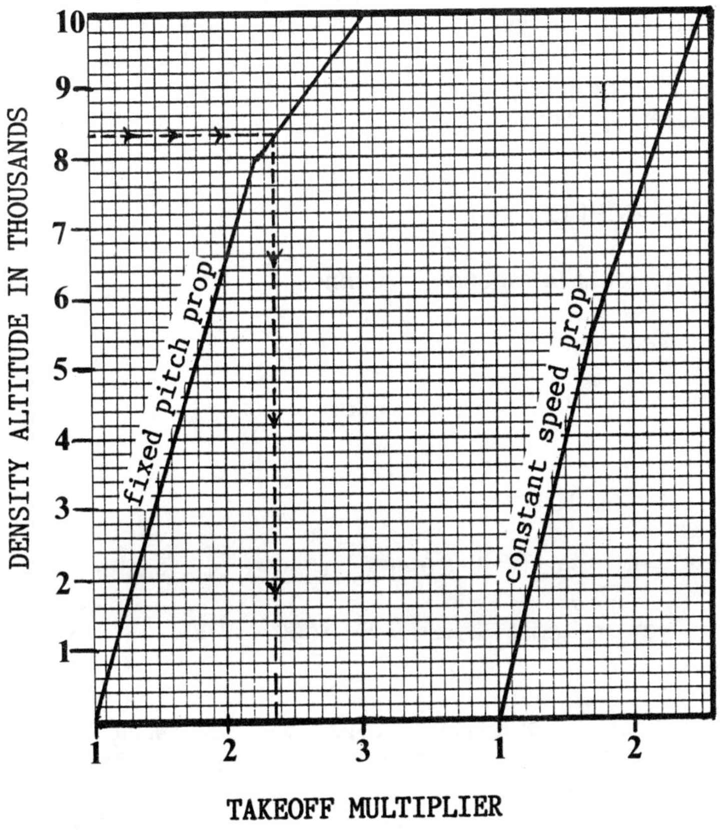

Multiply the sea level takeoff distance priviously obtained either from this book or other sources by the factor obtained from this graph. Note that both fixed pitch and constant speed information is presented.

Figure 1-9

WIND EFFECT

Rule Of Thumb #11 -- A headwind which is 10% of the takeoff airspeed will reduce the takeoff distance by 19%. A tailwind which is 10% of the takeoff airspeed will INCREASE the takeoff distance by 21%.

EXAMPLE:

Takeoff airspeed = 70 knots

Headwind force = 7 knots

No wind takeoff distance = 1000 feet

Ratio of wind to takeoff airspeed = 7 ÷ 70 = .10 or 10%

New takeoff distance = 1000 x .19 = 190

New takeoff distance = 1000 - 190 = 810 feet

Use the graph in Figure 1-10 to determine the wind effects on no wind takeoff distance. Obtain the ratio of wind speed to takeoff airspeed by dividing the wind speed by the takeoff airspeed in knots.

WIND EFFECT GRAPH

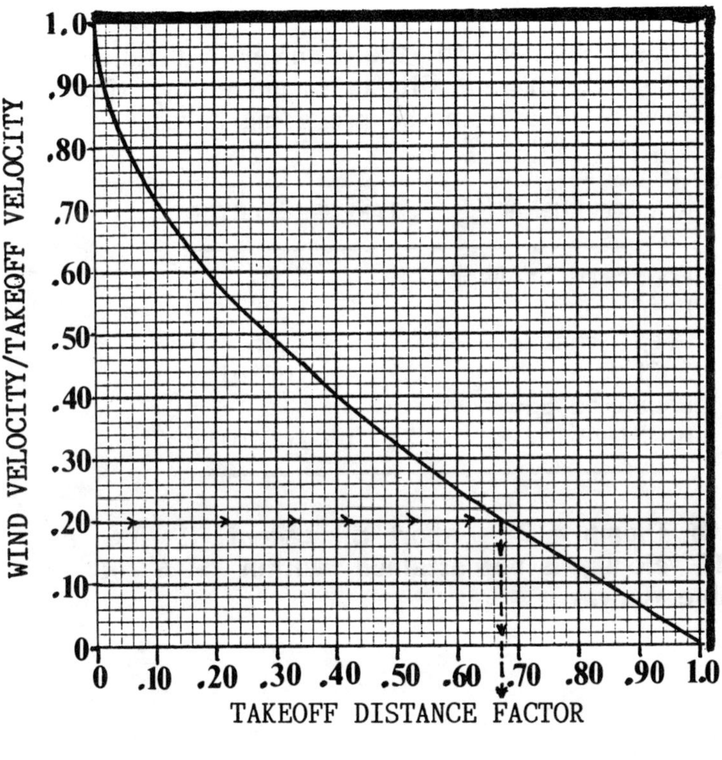

EXAMPLE:
Ratio of Wind Velocity to Takeoff Velocity = 20%. Takeoff Distance Factor = .67. Multiply Factor by Takeoff Distance obtained either from this book or other sources.

Figure 1-10

RUNWAY SURFACE EFFECT

Use the table below to determine the effects of various runway surfaces on takeoff distance.

Apply these figures to the runway distance required by the Airplane Flight Manual or by the calculations derived from the preceding pages.

RUNWAY SURFACE	TAKEOFF FACTOR
Concrete or Asphalt	0.0
Hard Turf	1.08
Deep Grass	1.25
Soft Field	1.25 to infinity
Unpacked Snow or Slush	2.0 to infinity

Rule Of Thumb #12 — Abort the takeoff if 70% of takeoff velocity is not attained within 50% of available runway.

CHAPTER 2

LIGHT TWIN AIRCRAFT

CHAPTER CONTENTS

RULES OF THUMB

FORMULAS

SPECIAL CONSIDERATIONS FOR LIGHT TWIN PILOTS

The Light Twin airplane presents special performance problems to the pilot. These airplanes have been called nothing more than single engine airplanes with their power divided into two packages. This statement does seem to have some validity in that many light twins have all engine power to weight ratios similar to single engine aircraft. The major advantage of having more than one power source is that it is possible to duplicate such things as alternators and instrument air pumps; thereby increasing the safety and flexibility of the aircraft. The major disadvantage of having multiple power sources is that the pilot must be able to very accurately plan and execute his takeoff to allow him to maintain positive aircraft control in the event of the sudden failure of one engine. It may be well worth your while to spend an evening with the information in this section and your Airplane Flight Manual.

Use the information found here in addition to the information found elsewhere in this book.

Rule Of Thumb #13 -- To determine Lift Off Speed, use the manufacutrer's recomended speed or Vmc +5 knots whichever is greater.

Rule Of Thumb #14 -- Initial climb speed should not be slower than the single-engine best-rate-of-climb speed (Vyse) and not faster than the all-engine best-rate-of-climb speed (Vy).

Rule Of Thumb #15 -- Twin engine aircraft weighing more than 6000 pounds and/or having a Vso in excess of 61 knots should be able to produce a minimum single engine climb rate of at least 100 feet per minute at 5000 feet I.S.A. with the inoperative engine feathered and the aircraft in clean configuration.

The reason for this is that the F.A.A. requires a minimum climb capability for these aircraft be demonstrated under the conditions stated in the rule of thumb. The climb performance required is determined by the formula:

$$\text{Rate of Climb} = 0.027 \times Vso^2$$

So:

An airplane having a Vso of 61 knots would have to be able to climb 100 feet per minute. This is so because:

$$\text{R.O.C.} = 61^2 \times 0.027 = 100.467$$

CAUTION:
THERE IS NO REQUIREMENT THAT THESE AIRCRAFT BE ABLE TO CLIMB AT SEA LEVEL IN THE TAKEOFF CONFIGURATION WITH ONE ENGINE INOPERATIVE.

Rule Of Thumb #16 -- The loss of 50% of the available thrust will result in a loss of 80% to 90% of the all engine rate of climb. The average loss is around 85%.

To determine the actual loss for your airplane use the following formula:

$$\text{Percent Loss} = 100-((\text{SROC}/\text{TROC}) \times 100)$$

WHERE:

SROC = sea level single engine rate of climb from aircraft manual.
TROC = sea level twin engine rate of climb from aircraft manual.

EXAMPLE:

Percent Loss = 100-((300/1200) x 100) = 75%

Leaving only 25% of the twin engine climb rate.

All airplanes require excess horsepower to climb, that is, power in excess of that required for straight and level flight. To determine the excess horsepower available for single engine climb use the following formula:

EHP = ((SROC X GROSS WEIGHT)/33,000)

WHERE:

SROC = Gross weight single engine rate of climb at sea level.
GROSS WEIGHT = Maximum certified gross weight.

EXAMPLE:

EHP = (300 x 4200)/33,000 = 38.18

Now that you know the single engine EHP for your airplane, you can determine the expected initial single engine rate of climb using the formula:

SROC = (EHP x 33,000)/ACTUAL WEIGHT

EXAMPLE:

SROC = (38.18 x 33,000)/3800 = 331.56

Now, using the information gained from this section you can apply the Density Altitude information in the following pages to determine the effect of non standard temperature on your initial climb capability.

CHAPTER 3

CLIMB PERFORMANCE

CHAPTER CONTENTS

CLIMB PERFORMANCE

While much of the information in this section is applicable to all airplanes, it may be of little value to more sophisticated aircraft as the manufacturer generally supplies complete performance information. Performance information is intended for use with non turbocharged reciprocating engine aircraft with a power loading of 10 to 15 pounds per horsepower.

In order to decide if the aircraft has the climb capability necessary under a given set of conditions, it will first be necessary to determine the desired climb profile.

Some departures and IFR missed approaches require a climb profile given in feet per mile. The graph in figure 3-1 allows you to convert this information into a rate of climb for the expected ground speed to make good the climb gradient required.

FORMULA:

Rate Required = F x (GS÷60)

Where:

F = required feet per mile in hundreds.

GS = estimated ground speed

BASIC LANGUAGE PROGRAM

```
10 INPUT"EST. G.S.=";S,"FT. PER MILE=";F
20 M=S/60:R=M*F:R=INT((10*R)+.05)/10
   :PRINT"REQD. RATE=";R;" F.P.M.":END
```

You may also use the graph in figure 3-1 to determine the rate required to climb over an obstacle of known height and distance. To do this, just divide the altitude in feet by the distance to get the required number of feet per mile and use the graph.

EXAMPLE:

Obstacle height above takeoff elevation = 3700 feet

Desired obstacle clearance = 1000 feet

Distance to obstacle = 10 miles

Estimated Ground speed = 120

Total height above takeoff elevation = 3700 + 1000 = 4700 feet

Feet per mile = 4700/10 = 470

Rate required = 940 feet per minute (from graph).

THE FORMULA FOR THIS IS:

$$ROC = \frac{GS \times (ALT/6076)}{60\sqrt{DIST^2 + (ALT/6076)^2}} \times 6076$$

Where:

GS = estimated groundspeed

ALT = elevation change

CLIMB PERFORMANCE

DIST = distance in nautical miles

BASIC LANGUAGE PROGRAM

```
10 INPUT"CLIMB G.S.=";T,"ALT. CHANGE=";
   A,"DIST.=";D
15 A=A/6076:R=(T*A)/(60*√((D↑2)+(A↑2)))
20 R=R*6076:PRINT"R.O.C. REQD=";R:END
```

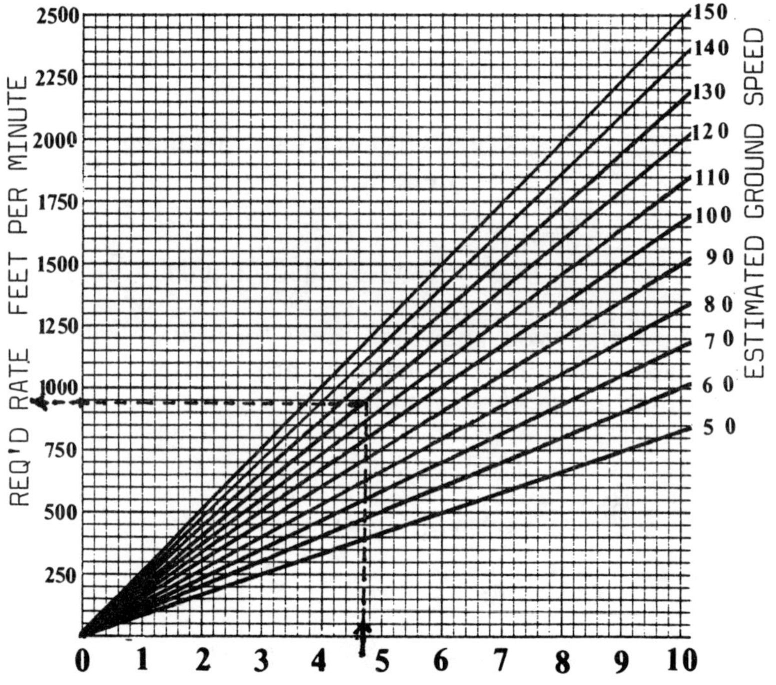

FEET PER MILE / HUNDREDS

EXAMPLE: Req'd 470 feet per mile. Expected climb ground speed is 120 knots.
Req'd rate is 940 feet per minute.

Figure 3-1

AVAILABLE CLIMB PERFORMANCE

All airplanes require excess horsepower to climb, that is, power in excess of that required to maintain straight and level flight. To determine the excess horsepower available for your aircraft use the following formula:

$$EHP = \frac{ROC \times GROSS\ WEIGHT}{33{,}000}$$

Where:

ROC = sea level gross weight rate of climb.
GROSS WEIGHT = maximum certificated gross weight.

EXAMPLE:

EHP = (792 x 2450)/33,000 = 58.8

Now that you know the EHP for your airplane, you can determine the approximate initial rate of climb using the formula:

$$ROC = \frac{EHP \times 33{,}000}{ACTUAL\ WEIGHT}$$

EXAMPLE:

ROC = (58.8 x 33,000)/2450 = 792

Another formula that seems to work well with these airplanes is:

$$ROC = \left(\frac{GROSS\ WEIGHT}{ACTUAL\ WEIGHT}\right)^2 \times S.L.ROC$$

Where:

S.L.ROC = Published sea level rate of climb at gross weight

EXAMPLE:

$(2450/2200)^2 \times 792 = 982.23$

So, the initial rate of climb for this aircraft should be around 982 feet per minute at a weight of 2200 pounds.

You will find this information in the graph in Figure 3-2 or the table in Figure 3-3. To use either of these, divide the actual weight by the gross weight and find the factor for this value, then multiply the sea level gross weight rate of climb by the factor. The answer will be your expected initial rate of climb.

Remember to use the sea level gross weight performance information as published in the Flight or Owners Manual for these computations.

Once you have determined the effect of weight on your climb capability move on to the Density Altitude section to determine the effect of temperature and altitude on the sea level climb performance.

Use the tables and graphs in the Takeoff Section to obtain Density Altitude values.

RATE OF CLIMB vs WEIGHT

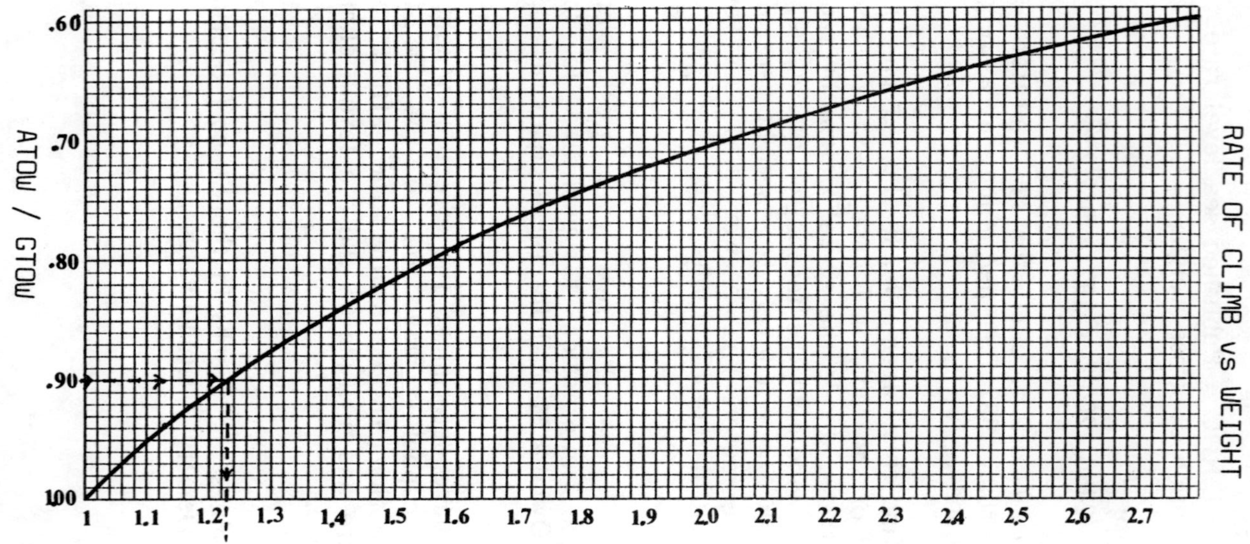

RATE OF CLIMB FACTOR

EXAMPLE: Actual Takeoff Weight divided by Gross Takeoff Weight = .90. Climb factor = 1.23 Multiply published sea level gross weight climb by climb factor to obtain expected climb.

Figure 3-2

RATE OF CLIMB vs WEIGHT
TABLE

ACT.WT./G.WT.	CLIMB FACTOR
1.00	1.00
.90	1.23
.80	1.56
.70	2.04
.60	2.77

To use this table divide the Actual Weight by the certificated Gross Weight then use this value to find the Climb Factor, then multiply the Climb Factor by the sea level gross weight rate of climb.

Figure 3-3

BASIC LANGUAGE PROGRAM

```
10 INPUT"CERT. GROSS WT.=";G,"ACT.
   GROSS WT.=";W,"S.L. GR. WT. ROC="
   ;R
20 E=((G/W)↑2)*R:E=INT((100*E)+.05)
   /100
30 PRINT"EXP. ROC=";E;" FT/MIN":END
```

This program gives the expected rate of climb for the actual weight of the airplane. Enter the sea level gross weight climb information from the Flight Manual or Operating Manual.

DENSITY ALTITUDE AND RATE OF CLIMB

Rule Of Thumb #17 — (fixed pitch props) Reduce Sea Level Rate of Climb 7% for each thousand feet Density Altitude up to 8500 feet and 8% for each thousand feet above 8500 feet.

Rule Of Thumb #18 — (variable pitch props) Reduce Sea Level Rate of Climb 6% for each thousand feet Density Altitude up to 8500 feet and 8% for each thousand feet above 8500 feet.

EXAMPLE: (fixed pitch)

Sea Level ROC = 1000 feet per min.

Density Altitude = 4000

4 x 7% = 28%

1000 x .28 = 280

1000 - 280 = 720

Expected Rate Of Climb = 720 f.p.m.

EXAMPLE: (variable pitch)

Sea Level ROC = 1000 feet per min.

Density Altitude = 4000

4 x 6% = 24%

1000 x .24 = 240

1000 - 240 = 760

Expected Rate Of Climb = 760 f.p.m.

CAUTION: For Density Altitudes above 8500 feet CHECK THE AIRCRAFT MANUAL.

Use the graph in Figure 3-4 to determine Density Altitude multiplier.

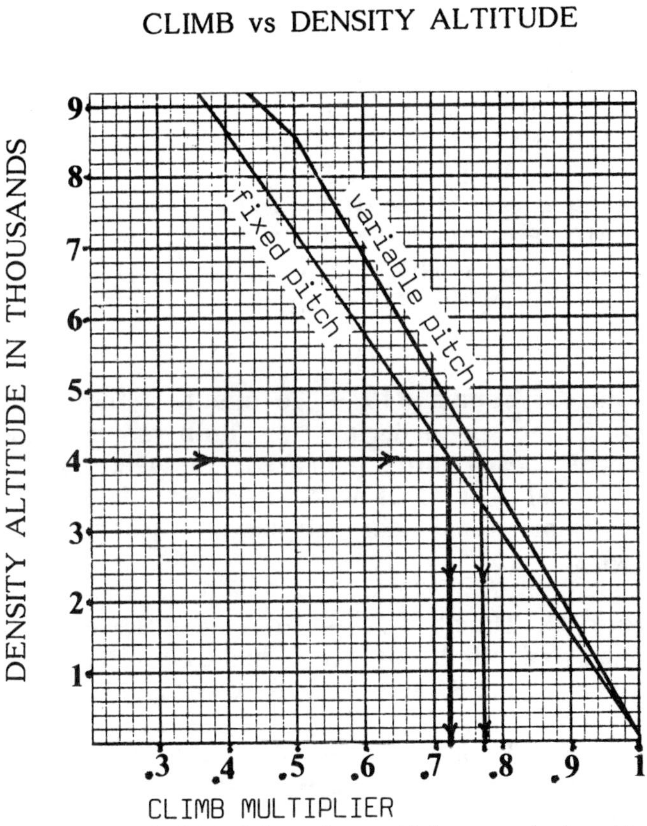

Multiply Climb Multiplier from this graph by the Sea Level Rate of Climb to get expected rate of climb. EXAMPLE:

Density Altitude = 4000. Climb Multiplier = .72 for fixed pitch or .77 for variable pitch.

Figure 3-4

BASIC LANGUAGE PROGRAM

```
10 INPUT"FIXED(F)   VARIABLE(V)";A$
15 INPUT"S.L. ROC=";R,"DEN. ALT.=";A
20 IFA$<>"F"ANDA$<>"V"THEN10
30 IFA$="F"THEN LET X=.00007:Y=.00008
   :GOTO100
40 X=.00005875:Y=.00008
100 NR=(1-(X*A))*R
110 IFA$="F"ANDA>8500THEN LET NR=(1-(.595
    +((A-8500)*Y)))*R
120 IFA$="V"ANDA>8500THEN LET NR=(1-(.499
    +((A-8500)*Y)))*R
200 NR=INT((NR*100)+.5)/100
210 IFA>8500THEN PRINT"ROC=";NR;
    "     *CHECK MANUAL*":END
220 PRINT"ROC=";NR:END
```

Rule Of Thumb #19 -- To maintain the Best Rate Of Climb speed during the climb reduce the Sea Level best rate indicated airspeed 1% per thousand feet. This value is about 1 knot per thousand for most airplanes through medium twins.

Rule Of Thumb #20 -- To reduce the effects of a headwind climb at the highest airspeed that gives an acceptable rate of climb.

Rule Of Thumb #21 -- To take advantage of a tailwind climb at the best rate of climb airspeed.

CHAPTER 4

WEIGHT AND BALANCE

CHAPTER CONTENTS

BASIC WEIGHT AND BALANCE FORMULA

Weight x Arm = Moment

This is the basic formula upon which all weight and balance calculations are based.

The arm can be found by adapting the formula:

$$\text{Arm} = \frac{\text{Moment}}{\text{Weights}}$$

To find the CG location use the formula:

$$\text{Arm} = \frac{\text{Total Moments}}{\text{Total Weights}}$$

UNIVERSAL BASIC LANGUAGE PROGRAM

```
10 CLEAR
20 DIMA(10):DIMB(10)
30 INPUT"EMPTY WEIGHT=";A(0),"EMPTY
   MOMENT=";BC
40 I=I+1
50 INPUT"WEIGHT=";A(I)
60 INPUT"ARM=";B(I)
70 C=C+(A(I)*B(I)):IF I=10THEN GOTO 200
80 WAIT0:PRINT"END=(E) CONTINUE=(C)"
90 A$=INKEY$:IFA$="E"THEN GOTO 200
100 IFA$<>"C"THEN GOTO 80
110 GOTO 40
200 WAIT:FOR J=0 TO I:A=A(J)+A:NEXTJ:C=BC+C
```

```
210 PRINT"TOTAL WT.=";A:PRINT"TOTAL MOM.=
    ";C:CG=C/A
220 PRINT"C.G.=";CG;" IN.":END
```

This program can accept 10 weights and 10 arms. If you need more than this, change the DIM statement in line 20 and the IF statement in line 70 to allow for more inputs.

To use the program enter the information requested, when End or Continue prompt appears press "C" to continue entering information. When all information is entered press "E" at the End or Continue prompt. This causes the program to compute and print out the Total Weight, Total Moment and Center of Gravity location.

WEIGHT SHIFT FORMULA

$$\frac{\text{weight shifted}}{\text{total weight}} = \frac{\text{distance CG shifted}}{\text{distance weight shifted}}$$

This formula can be utilized to shift weight if the CG is found to be out of limits. Use of this formula avoids working the entire problem over again by trial and error.

The following forms of this formula can be used to determine the different elements of the formula.

To find CG CHANGE when WEIGHT CHANGE AND DISTANCE are known:

$$\text{CG CHANGE} = \frac{\text{WS x DIST}}{\text{TW}}$$

Where:
WS = Weight Shifted
DIST = Distance Weight Shifted
TW = Total Weight

To find the WEIGHT TO SHIFT when CG CHANGE NECESSARY AND DISTANCE are known:

$$\text{WEIGHT SHIFTED} = \frac{\text{TW x CG CHANGE}}{\text{DIST}}$$

To find the DISTANCE TO SHIFT when CG CHANGE NECESSARY AND WEIGHT TO SHIFT are known:

$$\text{DISTANCE TO SHIFT} = \frac{\text{TW x CG CHANGE}}{\text{WS}}$$

BASIC LANGUAGE PROGRAM

```
10 INPUT"CG CNG(1)WT CNG(2)DIST(3)";A
20 ON A GOTO 30,60,90
30 INPUT"WT. SHIFTED=";WS,"DIST. SHIFTED=";D
   ,"TOTAL WT.=";TW
40 CG=(WS*D)/TW
50 PRINT"CG. CHNG=";CG:END
60 INPUT"TOTAL WT.=";TW,"CG SHIFT=";CG,"DIST
   SHIFT=";D
```

```
70 WS=(TW*CG)/D
80 PRINT"WEIGHT SHIFT=";WS:END
90 INPUT"TOTAL WT.=";TW,"CG SHIFT=";CG,
   "WEIGHT SHIFT=";WS
100 D=(TW*CG)/WS
110 PRINT"DISTANCE=";D:END
```

To use this program press the key (1, 2 or 3) for the program desired and enter the information requested. Pressing the "1" key will solve for the C.G. Change when the Weight Shifted, Distance Weight is to be shifted and Total Weight are known. Pressing the "2" key will solve for Weight to be Shifted when Total Weight, Distance to Shift C.G. and Distance Weight is to be Shifted are known. Pressing the "3" key will solve for Distance to Shift Weight when Total Weight, Distance to Shift C.G. and Weight Shifted are known.

Rule of Thumb #22 — To obtain the maximum True Airspeed and range, load the aircraft as close as possible to the aft C.G. limit for the aircraft weight.

This is because the drag is reduced due to a smaller angle of attack of the stabilizer and less downward force by the stabilizer is required to overcome the nose-down pitching tendency.

Rule of Thumb #23 — The airplane will stall at a higher speed with a forward C.G. location.

This is due to the fact that the stalling angle of attack is reached at a higher speed due to increased wing loading. This increased wing loading is created by the increased force required from the stabilizer to overcome the increased nose-down pitching tendency caused by the forward C.G. location.

Pilots must be very careful to determine that the aircraft will be within C.G. limits for **both takeoff and landing.** Some aircraft experience an aft C.G. movement as fuel is burned off and may exceed aft C.G. limit prior to landing.

One trick many Professional Pilots use is to determine in advance various loading conditions by working a number of commonly encountered weight and balance conditions, then recording this information for future reference. This short cuts the time required for takeoff planning. Once the takeoff weight and balance condition is determined it is a simple matter to subtract the estimated fuel burn and compute the landing weight and balance condition.

Since we can safely assume that the airplane will always be flown with a pilot and other items such as Charts, Survival Gear, Flashlights, etc.; these items and their moments may be added to the aircraft Basic Empty Weight and Moment to obtain the Basic Operating Condition. Using this figure eliminates the need to calculate for these items each time we do a weight and balance problem.

EXAMPLE:

Condition 1 (full fuel)

	weight	moment
Basic Condition	**1670**	**1837**

	weight	moment
Fuel	222	259
Totals	1892	2096

Minimum Moment=2055
Maximum Miment=2236

Condition 2 (1 passenger, full fuel)

	weight	moment
Basic Condition	1670	1837
2nd Seat Occupant	170	187
Fuel	222	259
Totals	2062	2283

Minimum Moment=2276
Maximum Moment=2437

The information in this section applies to all airplanes, no matter how large or how small; only the size of the numbers changes.

Following the advice offered here can save you time and money as well as greatly enhance the safety of your flight operations.

CHAPTER 5

AIRSPEED CALCULATIONS

CHAPTER CONTENTS

TRUE AIRSPEED FROM INICATED AIRSPEED

Rule Of Thumb #24 -- True airspeed increases 2% per thousand feet above sea level.

EXAMPLE:

IAS = 150

ALTITUDE = 5000

5 x 2% = 10%

150 x .10 = 15

TAS = 150 + 15 = 165

This rule is reasonably accurate where true airspeeds are in the 100 to 200 MPH range and the altitude is below 10,000 feet and the temperature is near ISA.

ACTUAL FORMULA:

$$TAS = IAS\sqrt{\frac{273.16 + C}{\left(\frac{288}{10^{D}}\right)}}$$

WHERE:

C = degrees Celsius

A = pressure altitude

$$D = \frac{A}{63691.776-(.2190731712\ A)}$$

BASIC LANGUAGE PROGRAM:

```
10 INPUT"IAS=";I,"PRESS. ALT.=";A,"TEMP
   C=";C
20 T=I*√((273.16+C)/(288/10↑(A/(63691.776
   -.2190731712*A))))
30 T=INT((100*T)+.05)/100:PRINT"TAS=";T:
   END
```

MACH TO TRUE AIRSPEED

See the tables in figure 2-1 for approximate TAS from MACH.

ACTUAL FORMULAS:

$$TAS = 38.967M\sqrt{273.16+C}$$

WHERE:

M = Mach number

C = True OAT Celsius

OR

If RAT recovery coefficient is known (usually .7 to 1.0).

THEN:

$$TAS = 38.967M\sqrt{\frac{273.16+C}{\left(\frac{1+ RM^2}{5}\right)}}$$

WHERE:

M = Mach number

C = ram air temperature

R = recovery coefficient (Ct)

BASIC LANGUAGE PROGRAMS

```
10 INPUT"MACH=";M,"TOAT=";C
20 T=38.967*M*√(273.16+C)
30 T=INT((T*100)+.05)/100
40 PRINT"KTAS=";T
```

IF RAT RECOVERY COEFFICIENT KNOWN

```
10 INPUT"MACH=";M,"RAT=";C
20 T=38.967*M*√((273.16+C)/(1+R*M↑2/5))
30 T=INT((T*100)+.05)/100
40 PRINT"KTAS=";T
```

Substitute the RAT recovery coefficient for your airplane for "R" in line 20.

If you fly different airplanes, then you might want to add this line:

```
15 INPUT"RECOVERY(CT)=";R
```

Don't change line 20 if line 15 is used.

Rule Of Thumb #25 — Multiply the Mach number by 5.7 then move the decimal point two places to the right to get TAS.

EXAMPLE:

MACH = .80

TAS = 5.7 x .80 = 4.56

TAS = 456 Knots

TRUE AIRSPEED TO MACH

ACTUAL FORMULA:

$$M = \frac{TAS}{38.967\sqrt{273.16+C}}$$

WHERE:

TAS = knots true airspeed

M = Mach number

C = true outside air temperature Celsius

BASIC LANGUAGE PROGRAM:

```
10 INPUT"KTAS=";T,"TOAT C=";C
20 A=38.967*√(273.16+C):M=T/A
30 M=((100*M)+.05)/100:PRINT"MACH=";M
        :END
```

EFFECT OF TEMPERATURE ON TAS

Rule Of Thumb #26 — If maintaining constant Mach cruise a 1 degree Celsius temperature increase will result in a 1 knot increase in true airspeed. A 1 degree Celsius decrease in temperature will result in a 1 knot decrease in true airspeed.

Rule Of Thumb #27 — If the temperature remains constant a .01 change in Mach number will cause a 6 knot change in the true airspeed.

MACH 1 AT TEMPERATURE

ACTUAL FORMULA:

$$\text{MACH 1} = 38.967\sqrt{273.16+C}$$

WHERE:

C = degrees Celsius

BASIC LANGUAGE PROGRAM

```
10 INPUT"TEMP C=";C:M=38.967*√(273.16+C):
   M=INT((100*M)+.05)/100
20 PRINT"MACH 1=";M;" KTS":END
```

MACH TO TRUE AIRSPEED TABLE (ISA)

ALT. FEET	APPROXIMATE KNOTS AT MACH NUMBER .4	.45	.5	.55	.6	.65	.7
0	265	298	331	364	397	430	463
1000	264	297	330	363	396	429	462
2000	263	296	329	361	394	427	460
3000	262	295	328	360	393	426	458
4000	261	294	327	359	392	424	457
5000	260	293	325	358	390	423	455
6000	259	292	324	356	389	421	454
7000	258	291	323	355	387	420	452
8000	257	289	322	354	386	418	450
9000	256	288	321	353	385	417	449
10000	255	287	320	351	383	415	447
12000	254	285	317	349	380	412	444
14000	252	283	315	346	377	409	440
16000	250	281	312	343	374	406	437
18000	248	279	310	340	371	403	433
20000	246	277	308	338	369	400	430
22000	244	275	305	336	366	397	427
24000	242	272	303	333	363	393	424
26000	240	270	300	330	360	390	420
28000	238	268	298	327	357	387	417
30000	236	266	295	325	354	384	413
32000	234	263	293	322	351	380	409
34000	232	261	290	319	348	377	406
36000 and above	230	259	288	316	345	374	402

Figure 5-1

MACH TO TRUE AIRSPEED TABLE (ISA)
CONTINUED

APPROXIMATE KNOTS AT MACH NUMBER

ALT. FEET	.75	.8	.85	.9	.95	1
0	496	529	562	596	629	662
1000	495	528	561	594	627	660
2000	493	526	558	591	624	657
3000	491	524	557	589	622	655
4000	490	522	555	587	620	653
5000	488	520	553	585	618	650
6000	486	518	551	583	616	648
7000	485	516	549	581	614	646
8000	483	515	547	579	611	643
9000	481	513	545	577	609	641
10000	479	511	543	575	607	639
12000	476	507	539	570	602	634
14000	472	503	535	566	598	629
16000	468	499	530	562	593	624
18000	464	495	526	557	588	619
20000	461	492	522	553	584	615
22000	458	488	518	549	580	610
24000	454	484	514	545	575	605
26000	450	480	510	540	570	600
28000	446	476	506	536	565	595
30000	443	472	502	531	561	590
32000	439	468	497	526	556	585
34000	435	464	493	522	551	580
36000 and above	431	460	489	517	546	575

Figure 5-1

PILOT NOTES

CHAPTER 6

TIME, SPEED, DISTANCE, FUEL CONSUMPTION, SPECIFIC RANGE AND SPECIFIC ENDURANCE CALCULATIONS

CHAPTER CONTENTS

TIME, SPEED, DISTANCE AND FUEL

It is often necessary to know your speed in Miles per Minute. In most cases the following two rules will give acceptable results.

Rule Of Thumb (Mach) #28 -- To find the approximate Miles per Minute, use Indicated Mach Number and move the decimal one position right.

EXAMPLE:

Indicated Mach = .80

Miles per Minute = 8.0

Rule Of Thumb (TAS) #29 — To find the approximate Miles per Minute, round off TAS to nearest ten, drop the zero and divide by 6.

EXAMPLE:

True Airspeed = 238

Rounding TAS = 240

Drop Zero and Divide = 24 ÷ 6 = 4

Miles per Minute = 4.0

Making proper allowances for wind will help improve the accuracy of these calculations.

If Ground Speed is known, use Ground Speed in place of TAS in the second Rule. This will make the accuracy nearly 100%.

The most accurate results will occur when the formula below is applied.

FORMULA:

Ground Speed ÷ 60 = Miles per Minute

BASIC TIME, SPEED, DISTANCE FORMULAS:

T = D ÷ S

S = D ÷ T

D = S x T

WHERE:

T = Time in hours and tenths of hours or minutes converted to tenths of hours. (.1 hour = 6 minutes)

S = Speed (either KTS or MPH)

D = Distance (either nautical or statute)

EXAMPLE 1:

Find Time when Speed and Distance are known.

Conditions:

Estimated Ground Speed = 300 knots

Distance = 620 nautical miles

TIME = DISTANCE ÷ SPEED

TIME = 620 ÷ 300 = 2.07

TIME = 2.07 hours

EXAMPLE 2:

Find Speed when Distance and Time are known.

Conditions:

Distance Flown = 30 nautical miles

Time to Fly Distance = 5.5 minutes

SPEED = DISTANCE ÷ TIME

SPEED = 30 ÷ 5.5 = 5.45 miles per minute

Note: distance divided by time in minutes gives miles per minute.

SPEED = 5.45 x 60 = 327

SPEED = 327 knots

EXAMPLE 3:

Find Distance when Speed and Time are known.

Conditions:

Speed = 300 knots

Time Flown = .2 hours

DISTANCE = SPEED x TIME (hours and tenths)

DISTANCE = 300 x .2 = 60

DISTANCE = 60 nautical miles

DON'T FORGET -- do not attempt to mix statute miles with speeds in knots or vice versa.

BASIC LANGUAGE PROGRAM

```
10 INPUT"SPEED(1) DIST.(2) TIME(3)";A
15 IF A<>1ANDA<>2ANDA<>3THEN10
20 ON A GOTO 50,60,70
50 INPUT"DISTANCE=";D,"TIME(HRS.TENTHS)=";T
52 S=D/T
55 PRINT"SPEED=";INT((100*S)+.05)/100:END
60 INPUT"SPEED=";S,"TIME(HRS.TENTHS)=";T
62 D=S*T
65 PRINT"DISTANCE=";INT((D*100)+.05)/100
   :END
70 INPUT"DISTANCE=";D,"SPEED=";S
72 T=D/S
75 PRINT"TIME(HRS.TENTHS)=";INT((100*T)+
   .05)/100:END
```

To use this program select the number corresponding to the value you want to solve for and enter the requested information. Pressing the "1" key will give the Speed when the time and distance are known. Pressing the "2" key will give the Distance when speed and time are known. Pressing the "3" key will give the Time when speed and distance are known.

REMINDER - When using this program do not mix knots and statute values. If you use nautical terms then the answers will be in nautical values. If you use statute terms then the answers will be in statute values.

FUEL CONSUMPTION

BASIC FORMULAS:

$$GPH = \frac{GALLONS}{HOURS}$$

$$MPG = \frac{GALLONS}{MILES}$$

$$TIME\ REMAINING = \frac{FUEL\ REMAINING}{FUEL\ RATE}$$

Substituting Pounds or Liters for Gallons in the above formulas will give fuel consumption in Pounds or Liters per Hour or Miles per Pound or Liter. Be careful not to mix Gallons, Liters or Pounds; use the same value throughout your calculations.

BASIC LANGUAGE PROGRAM

```
10 INPUT"GPH(1)MPG(2)TIME(3);A
20 IFA<>1ANDA<>2ANDA<>3THEN10
30 ON A GOTO50,60,70
50 INPUT"GAL,LBS,ETC=";U,"TIME(HRS.TENTHS)
   =";T
52 R=U/T:R=INT((R*100)+.05)/100
55 PRINTR;" PER HOUR":END
60 INPUT"MILES FLOWN=";D,"FUEL USED(GAL.
   LBS.ETC)=";UU
62 PU=D/UU:PU=INT((PU*100)+.05)/100
65 PRINTPU;" MILES PER UNIT":END
70 INPUT"FUEL REMAINING(GAL,#,ETC)=";UR,
   "FUEL RATE(GAL,#,ETC)=";R
```

```
72 T=UR/R:T=INT((T*100)+.05)/100
75 PRINTT;" HRS REMAINING":END
```

To use this program select the number corresponding to the value you wish to solve for and enter the requested information. Pressing the "1" key will give Gallons Per Hour, Pounds Per Hour, Liters Per Hour, etc. (depending upon which unit of measurement is entered into the program) when the Unit of Measurement and Time are known. Pressing the "2" key will give Miles per Unit when Miles Flown and Fuel Used are known. Pressing the "3" key will give Time Remaining when Fuel Units Remaining and Fuel Rate are known.

For ease in calculation convert time to hours and tenths of hours. To convert hours and minutes to hours and tenths divide the minutes by 6.

EXAMPLES:

1 hour 24 minutes = 1.4 hours

30 minutes = .5 hours

ACTUAL FORMULA:

$$\frac{(\text{Hours} \times 3600) + (\text{Minutes} \times 60)}{3600}$$

Sometimes fuel calculations are made using either "specific range" or "specific endurance". Specific Range is the number of nautical miles of flying distance per pound of fuel. Specific Endurance is the number of hours flying time per pound of fuel.

SPECIFIC RANGE FORMULA

$$\text{specific range} = \frac{\text{knots}}{\text{fuel flow}}$$

OR

$$\text{specific range} = \frac{\text{nautical miles}}{\text{lbs of fuel}}$$

OR

$$\text{specific range} = \frac{\text{nmi. per hr.}}{\text{lbs. fuel per hr.}}$$

EXAMPLE:

Nmi. per hr.(knots) = 300

Lbs. fuel per hour = 1200

Specific range = 300 1200

Specific range = .25

This means that the aircraft used in this example will fly .25 nautical miles for each pound of fuel consumed. Or you could say the aircraft will consume 4 pounds of fuel for each nautical mile flown.

You should be aware that wind will have an effect on specific range unless the speed used in the formula is the ground speed being flown.

BASIC LANGUAGE PROGRAM

```
10 INPUT"NAUTICAL MILES=";D,"FUEL POUNDS
   =";F
20 R=D/F:R=INT((100*R)+.05)/100
30 PRINT"SPEC. RNG.=";R
```

This program gives the Specific range when the elements of any one of the above formulas are entered in the computer. If total miles and total pounds are entered, the answer will be specific range for the entire trip. If ground speed and pounds per hour are entered, the answer will be specific range for present flight conditions.

SPECIFIC ENDURANCE FORMULA

$$\text{specific endurance} = \frac{1}{\text{fuel flow}}$$

OR

$$\text{specific endurance} = \frac{\text{flight hours}}{\text{lbs. of fuel}}$$

OR

$$\text{specific endurance} = \frac{\text{flight hours}}{\text{lbs. fuel per hr.}}$$

EXAMPLE:

Total flight hours = 1.7

Total fuel used = 2000 pounds

Specific endurance = 1.7 2000

Specific endurance = .00085 hours per lb.

The aircraft used in this example will fly .00085 hours per pound of fuel consumed provided the temperature remains constant and no power changes are made.

BASIC LANGUAGE PROGRAM

```
10 INPUT"HOURS(HRS.TENTHS)=";T,"FUEL
   POUNDS=";F
20 E=T/F
30 PRINT"SPEC. ENDUR.=";E
```

This program gives the specific endurance when hours and tenths and fuel pounds are entered. Use the elements of any one of the formulas above. If total time and total fuel in pounds are entered the answer will be the specific endurance for the entire trip. If time and the amount of fuel used for that time are entered, then the answer will be specific endurance for that portion of the flight.

The specific range and specific endurance may be used for inflight calculations, or for flight planning, if you desire. I have found that information gained from these formulas is most useful in establishing average fuel consumption and average expected endurance.

PILOT NOTES

CHAPTER 7

CRUISE CONTROL

CHAPTER CONTENTS

RULES OF THUMB

GRAPHS

BASIC LANGUAGE PROGRAMS

NON TURBOCHARGED RECIPROCATING AIRCRAFT

Rule Of Thumb #30 -- The maximum cruise speed for a given power setting will occur at the maximum density altitude where that power setting can be maintained at full throttle. Therefore, for maximum speed fly the nearest legal altitude to that required to give the desired cruise condition. Use the following Rule Of Thumb or the Graph in Figure 7-1 to determine the approximate percentage of horsepower available at altitude.

Rule Of Thumb #31 — Available full throttle horsepower decreases approximately 3% for each thousand feet density altitude above sea level.

EXAMPLE:

Cruise Density Altitude = 6500

Percent H.P. = 6.5 x .03 = .195

Percent H.P. = 1 - .195 = .805

Percent H.P. = 80.5%

BASIC LANGUAGE PROGRAM

```
10 INPUT"S.L.HRSPWR=";H,"DEN. ALT.=";A
20 D=1-(A*.00003):P=H-(D*H)
30 PRINT"% H.P.=";D
40 PRINT"H.P.=";P:END
```

NOTE: Use the Graph in Figure 1-6 or the Tables in Figure 1-7 to determine the Density Altitude for the preceding calculations.

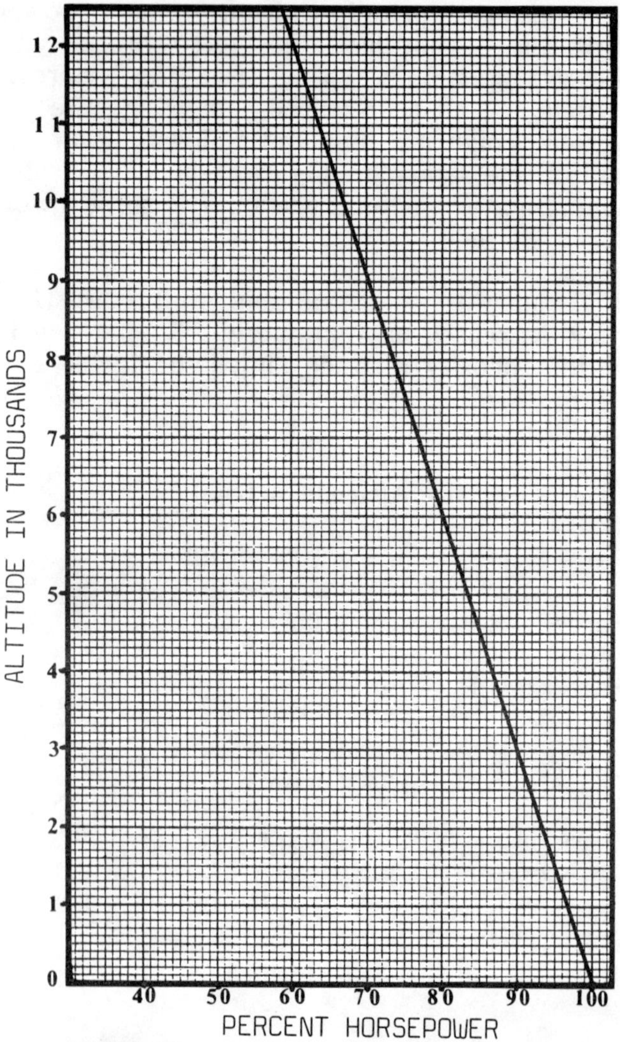

EXAMPLE:
Density altitude = 10,000 feet
Percent horsepower = 67%

Figure 7-1

Rule Of Thumb #32 -- To determine the approximate best Indicated Airspeed for greatest range, multiply the clean power-off stall speed (bottom of the green arc) by one of the factors below:

Single engine fixed gear = 1.6

Single engine retractable = 2.0

Light twin = 1.7

Rule Of Thumb #33 -- To maintain the maximum range Indicated Airspeed reduce the Indicated Airspeed by 10% for each 20% decrease in aircraft weight. This will require that you reduce the power periodically to maintain the best speed.

TURBOPROP AND TURBINE AIRPLANES

Rule Of Thumb #34 -- The shortest trip time may be obtained by using an altitude that is equal to the distance plus the departure field elevation in hundreds of feet, the maximum altitude is the altitude for maximum True Airspeed.

EXAMPLE:

Trip dist. = 175 nmi.

Field elevation = 3500

Flight level = 175 + 35 = 210

Flight level = 210

Rule Of Thumb #35 — The minimum fuel used may be obtained by using an altitude that is equal to 1.5 times the distance plus the departure field elevation in hundreds of feet, the maximum altitude is the best range altitude or maximum certificated altitude.

EXAMPLE:

Trip dist. = 200 nmi.

Field elevation = 1000

Flight level = 200 x 1.5 = 300

Flight level = 300 + 10 = 310

Flight level = 310

Rule Of Thumb #36 — If step climb is required, climb to the next usable flight level when the indicated cruise airspeed increases 5 knots. If above FL 290 allow the cruise airspeed to increase 10 knots before attempting climb.

ENROUTE WIND EFFECT

Rule Of Thumb #37 — Increase maximum range speed by 10% when flying into a headwind and reduce maximum range speed by 5% when flying with a tailwind. This rule works pretty well with all airplanes and it will help maintain a lower overall average operating cost.

ENDURANCE

Rule Of Thumb (non turbocharged reciprocating aircraft) #38 -- The Maximum Endurance is obtained at sea level. The reason for this is that it takes less power to maintain the same indicated airspeed at sea level than it does at altitude.

Rule Of Thumb #39 -- To find the Maximum Endurance speed multiply the clean power-off stall speed (bottom of the green arc) by:

1.2 for single engine fixed gear.

OR

1.3 for retractable singles and light twins.

NOTE: The Maximum Endurance airspeed WILL NOT produce maximum range.

PILOT NOTES

CHAPTER 8

MANEUVERING FLIGHT

CHAPTER CONTENTS

RULES OF THUMB

TABLES

GRAPHS

FORMULAS

BASIC LANGUAGE PROGRAMS

TURNS

Rule Of Thumb (MPH) #40 — To determine the angle of bank required to make a stantard rate turn at various airspeeds in miles per hour, divide the True Airspeed by 10 and add 5.

EXAMPLE:

TAS = 170 MPH

Bank Angle = 170 ÷ 10 = 17

Bank Angle = 17 + 5 = 22

Bank Angle = 22 degrees

BANK ANGLE vs TRUE AIRSPEED

TAS MPH	BANK ANGLE	TAS MPH	BANK ANGLE
60	8	160	21
70	10	170	22
80	11	180	23
90	12	190	24
100	14	200	26
110	15	210	27
120	16	220	28
130	17	230	29
140	19	240	30
150	20	250	31

Figure 8-1

FORMULA:

$$\text{BANK ANGLE} = \text{ATN}\left(\frac{9.2177478V}{3860}\right)$$

Where:

V = velocity in MPH

BASIC LANGUAGE PROGRAM (MPH)

```
10 INPUT"TAS=";V:A=ATN((9.2177478*V)/
   3860)
20 A=INT((100*A)+.05)/100
30 PRINT"BANK ANGLE="A:END
```

Rule Of Thumb (Knots) #41 -- To determine the angle of bank for standard rate turn at various airspeeds in Knots, divide the True Airspeed by 10 and add one-half of the answer to the answer.

EXAMPLE:

TAS = 170 Knots

Bank Angle = 170 ÷ 10 = 17

Bank Angle = 17 + (17 2) = 17 + 8.5

Bank Angle = 25.5

Bank Angle = 25.5 degrees

BANK ANGLE vs TRUE AIRSPEED

TAS KTS	BANK ANGLE	TAS KTS	BANK ANGLE
60	9	160	24
70	11	170	25
80	12	180	26
90	14	190	28
100	15	200	29
110	17	210	30
120	18	220	31
130	20	230	32
140	21	240	33
150	22	250	34

Figure 8-2

FORMULA:

$$\text{BANK ANGLE} = \text{ATN}\left[1.15\left(\frac{9.2177478V}{3860}\right)\right]$$

Where:

V = velocity in KNOTS

BASIC LANGUAGE PROGRAM (KNOTS)

```
10 INPUT"TAS=";V:A=ATN(((9.2177478*V)/
   3860)*1.15)
20 A=INT((100*A)+.05)/100
30 PRINT"BANK ANGLE=";A:END
```

RADIUS OF TURN

Rule Of Thumb #42 — The radius of turn in miles is approximately equal to 1% of the True Airspeed or Ground Speed divided by 2.

EXAMPLE:

TAS or GS = 100 knots

Radius = 100 2 = 50

Radius = 50 x .01 = .50

Radius = .5 nautical miles

RADIUS OF TURN vs TRUE AIRSPEED

TAS	RADIUS	TAS	RADIUS
60	.31	160	.84
70	.37	170	.90
80	.42	180	.95
90	.47	190	1.00
100	.53	200	1.06
110	.58	210	1.11
120	.63	220	1.16
130	.69	230	1.22
140	.74	240	1.27
150	.79	250	1.32

Figure 8-3

When using the preceding Rule Of Thumb or the table in 8-3 the turn radius will be in nautical miles if knots are used, and in statute miles if miles per hour are used.

FORMULA:

$$\text{RADIUS} = \frac{(V/30)}{2\pi}$$

Where:

V = velocity MPH or KNOTS

BASIC LANGUAGE PROGRAM

```
10 INPUT"TAS=";V:R=(V/30)/(2*π)
20 R=INT((100*R)+.05)/100
30 PRINT"RADIUS=";R;" MI":END
```

TIMED TURNS

Rule Of Thumb #43 — To determine the number of seconds required to make the desired heading change at a standard rate, divide the number of degrees to be turned by 3 (number of degrees per second in standard rate turn). Multiply the answer by 2 if using a half-standard rate turn. Using this rule is quite useful when making turns on the magnetic compass.

EXAMPLE:

Degrees to Turn = 90

Time to turn = 90 ÷ 3 = 30

Time to turn = 30 seconds

If half-rate:

Time to turn = 30 x 2 = 60

Time to turn = 60 seconds

LOAD FACTORS

AIRPLANE CATEGORY (minimum required)

Category	Positive G	Negative G
Normal over 4000 pounds	2.5	-1.0
Normal	3.8	-1.52
Utility	4.4	-1.76
Acrobatic	6.0	-3.0

Figure 8-4

TURN DYMANICS

Use the graph in Figure 8-5 to determine the increase in load factor and stalling speed due to bank angle. Remember that these values cannot be achieved except by maintaining altitude while turning.

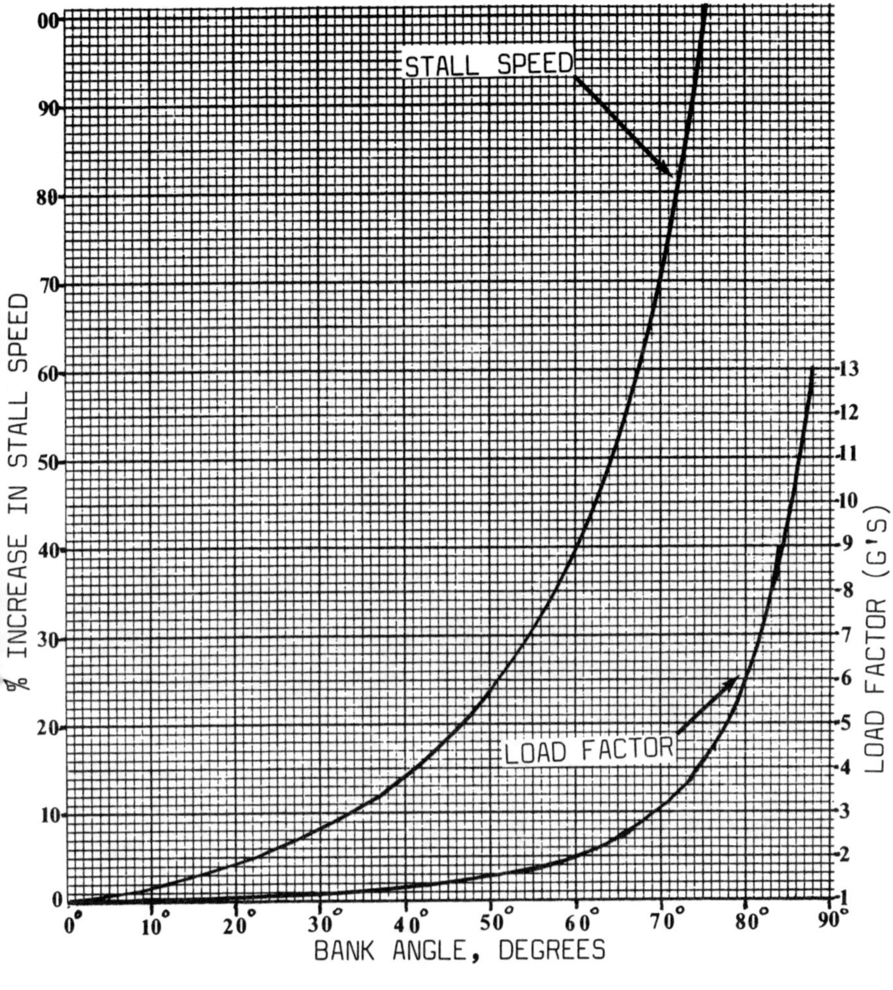

STALL SPEED AND LOAD FACTOR vs BANK ANGLE

Figure 8-5

Use the following formula to compute the G-force increase due to bank angle:

FORMULA:

$$G = \frac{1}{\text{COS B}}$$

Where:

G = G-force produced

B = bank angle in degrees

BASIC LANGUAGE PROGRAM

```
10 INPUT"BANK=";B
15 IF B>89.99999THENPRINT"INFINITY":END
20 G=1/COS B
30 G=INT((100*G)+.05)/100
40 PRINT"G FORCE=";G:END
```

Use the following formula to compute the increase in stalling speed due to bank angle:

FORMULA:

$$\text{Stall} = \frac{\text{Vs1}}{\text{COS B}}$$

Where:

Stall = stall speed at bank angle B

B = bank angle in degrees

Vs1 = wings level stall speed

BASIC LANGUAGE PROGRAM

```
10 INPUT"BANK=";B,"LVL STALL=";L
20 IF B>89.99999THENPRINT"INFINITY":END
30 S=L/ COSB
40 S=INT((100*S)+.05)/100
50 PRINT"STALL=";S:END
```

Use the following formula to compute the diameter of turn at Bank Angle and TAS:

FORMULA:

$$\text{Diameter} = \frac{\text{TAS}^2}{34208 \text{ TAN } B}$$

Where:

TAS = true airspeed in knots

B = bank angle in degrees

Diameter = diameter in nautical mi.

BASIC LANGUAGE PROGRAM

```
10 INPUT"TAS=";S,"BANK=";B
20 D=(S↑2)/(34208*TANB)
30 D=INT((D*100)+.05)/100
40 PRINT"DIAMETER=";D:END
```

Use the following formula to compute time required to turn 360 degrees at Bank Angle and TAS:

FORMULA:

$$\text{TIME} = \frac{0.0055\ \text{TAS}}{\text{TAN B}}$$

Where:

TIME = time in minutes

TAS = true airspeed in knots

B = bank angle in degrees

BASIC LANGUAGE PROGRAM

```
10 INPUT"TAS=";S,"BANK=";B
20 T=(.0055*S)/TANB
30 T=INT((100*T)+.05)/100
40 PRINT"TIME=";T;" MIN.":END
```

To compute the increase in stall speed due to G-force use the formula:

FORMULA:

$$\text{STALL} = \text{Vs1}\sqrt{\text{G}}$$

Where:

STALL = stall speed at G-force

G = G-force

Vs1 = wings level stall speed

BASIC LANGUAGE PROGRAM

```
10 INPUT"G-FORCE=";G,"VS1=";S
20 L=S*√G
30 L=INT((100*L)+.05)/100
40 PRINT"STALL=";L:END
```

MANEUVERING SPEED

This last formula or program can be used to compute the Maneuvering Speed (Va) for your airplane. To do this enter the limit load factor for your airplane as the G-force in the formula or program. This figure will be the Maneuvering Speed for your airplane at Gross Weight.

The Maneuvering Speed decreases as the square root of the weight decrease. Remember, as the gross weight decreases the Maneuvering Speed decreases.

The Maneuvering Speed for less than gross weight can be computed using the formula:

FORMULA:

$$Va = Va1 \quad \frac{AW}{GW}$$

Where:

Va = actual maneuvering speed

Va1 = gross weight maneuvering speed

AW = actual weight

GW = certificated gross weight

BASIC LANGUAGE PROGRAM

```
10 INPUT"ACT. WT.=";AW,"GR. WT.=";GW,
   "G.W. VA=";V
20 VA=V*  (AW/GW)
30 VA=INT((100*VA)+.05)/100
40 PRINT"VA=";VA:END
```

If you have an older General Aviation airplane you may not find a figure for Va in the Manual. If you have this problem then use the following Rule Of Thumb:

Rule Of Thumb #44 — For older airplanes use 1.7 Vs1 for the gross weight Maneuvering Speed.

EXAMPLE:

Clean stall speed = 60

Va = 60 x 1.7 = 102

Va = 102

MISCELLANEOUS

FORMULA FOR PIVOTAL ALTITUDE:

$$\text{Pivotal Altitude} = \frac{\text{TAS(mph)}^2}{15}$$

CHAPTER 9

NAVIGATION

CHAPTER CONTENTS

NAVIGATION

The first step in any navigation problem is to know where we are and where we want to go. Then, we need to find what course to fly, and the distance to the destination. For most of us this means turning to charts, plotters, tape, strings, pencils, bi-focals, rulers and a couple belts of Scotch. For those of you who prefer the latter or do not posses a programable calculator or computer, skip the first few pages of this chapter.

For those of you still with me, you may want to get a whole bottle of your favorite hooch before taking the dive into what is to come.

The following sets of formulas and programs will give you the initial course and distance between two points when you enter the latitude and longitude for each point. One will even give you the latitude of the point where the course line intercepts an intermediate longitude.

The first formula is the one the F.A.A. uses to establish all the airways in the Continental United States. I found this formula in AC 90-45A Appendix J dated 2/21/75. Those of you with an interest in this kind of mathematics will find a wealth of information in this AC.

At first glance this formula looks like the characters carved on the wall of an Egyptian Tomb, but take heart; all that Greek is only used to identify elements of the formula.

Now, even better news. I've developed a basic language program from this formula and you will find it following the pages containing the formula.

GREAT CIRCLE FORMULA #1:

Note: All angles are expressed in degrees and distances in nautical miles.

Given: B_1, L_1 Geodesic Lat., Lng. of P_1

B_2, L_2 Geodesic Lat., Lng. of P_2

$F = 3.3901 \times 10^{-3}$

$A_o = 3443.95594$ (semimajor axis of Earth)

Required: A_{12}: Azimuth from P_1 to P_2

S: Geodesic Length (distance)

Compute S (Geodesic Length in Nautical Miles):

$$S = (1-F)\, A_o\, S_1$$

Where:

$$S_1 = \Psi + \frac{(F^2+F)}{2}\left[\Psi(2-M) + \alpha\,(2A - M\phi)\right]$$

$$+ \frac{F^2}{16\alpha}\left\{8\Psi^2(M-1)\left[A + \alpha \operatorname{Cot}\frac{180\Psi}{\pi}\right] + \alpha^2\phi\right.$$

$$\left.\left[8A(M\phi - A) + (1-2\phi^2)\right] + \alpha\Psi\right\}$$

$$\Psi = \frac{\pi}{180} \operatorname{Arc\,Tan} \frac{\alpha}{\phi}$$

$$M = \frac{\alpha^2 - \text{Sin}^2 L \ \text{Cos}^2 \beta_1 \ \text{Cos}^2 \beta_2}{\alpha^2}$$

$$\alpha = \left[\text{Sin}^2 L \ \text{Cos}^2 \beta_2 + (\text{Sin} \beta_2 \ \text{Cos} \beta_1 - \text{Sin} \beta_1 \ \text{Cos} \beta_2 \ \text{Cos} L)^2 \right]^{1/2}$$

$$\phi = \text{Sin} \beta_1 \ \text{Sin} \beta_2 + \text{Cos} \beta_1 \ \text{Cos} \beta_2 \ \text{Cos} L$$

$$L = \begin{cases} L' \text{ if} \left| L' \right| \leq 180^\circ \\ L' + 360 \dfrac{\left| L' \right|}{L'} \text{ if} \left| L' \right| > 180^\circ \end{cases}$$

$$L' = L_1 - L_2$$

$$A = \text{Sin} \beta_1 \ \text{Sin} \beta_2$$

$$\beta_i = \text{Arc Tan} \left[(1-F) \ \text{Tan} \ B_i \right] \text{ for } i = 1, 2$$

Compute A_{12} (Azimuth from P_1 to P_2)

$$A_{12} = \begin{cases} \sigma \text{ if } L \geq 0 \\ \sigma + 180^\circ \text{ if } L < 0 \end{cases}$$

Where:

$$\sigma = \begin{cases} \text{Arc Tan}\ \frac{a}{b} + 180^{0}\ \text{if Arc Tan}\ \frac{a}{b} < 0 \\ \text{Arc Tan}\ \frac{a}{b}\ \text{if Arc Tan}\ \frac{a}{b} \geq 0 \end{cases}$$

$$a = \text{Sin H Cos}\ \beta_2$$

$$b = \text{Sin}\,\beta_2\ \text{Cos}\ \beta_1 - \text{Cos H Sin}\ \beta_1\ \text{Cos}\,\beta_2$$

$$H = \frac{180}{\pi}\left[\frac{\text{Sin L Cos}\ \beta_1\ \text{Cos}\ \beta_2}{\alpha}\right]\left[x\right] + L$$

$$x = \Psi(F+F^2) - \frac{AF^2}{2\alpha}\left[\alpha^2 + 2\ \Psi^2\right]$$

$$+ \frac{MF^2}{4}\left[\alpha\phi - 5\Psi + 4\Psi^2\ \text{Cot}\ \frac{180\Psi}{\pi}\right]$$

The Basic Language Program for this begins on page 121.

GREAT CIRCLE NAVIGATION
BASIC LANGUAGE PROGRAM #1:

```
7 INPUT"DEP. IDENT= ";Z$:GOTO 10
8 INPUT"DES. IDENT= ";Y$:RETURN
10 INPUT"LAT:DEP? ";K:INPUT"LNG:DEP? ";M
   :GOSUB 8:INPUT"LAT:DES? ";L:INPUT"LNG:
   DES? ";N
11 GOSUB60:XD=K:GOSUB100:K=ZT:XD=M:GOSUB100:
   M=ZT:XD=L:GOSUB100:L=ZT:XD=N
12 GOSUB100:N=ZT
13 C=.0033901:O=3443.95594
14 A=ATN((1-C)*TANK):B=ATN((1-C)*TANL):
   D=SINA*SINB
15 E=M-N:F=(SINA*SINB)+(COSA*COSB*COSE)
17 S=(((SINABSE)↑2)*((COSB)↑2))
20 T=(ABS(SINB*COSA-SINA*COSB*COSE)↑2):
   H=(S+T)↑.5
22 I=((H↑2)-((SINABSE)↑2)*((COSA)↑2)*((COSB
   )↑2))/H↑2:J=π/180*(ATN(H/F))
25 G=J+(((C↑2)+C)/2)*(J*(2-I)+(H*((2*D)
   -(I*C))))
27 V=(D+H*(1/(TAN((180*J)/π))))+((H↑2)*F)
   *((8*D)*((I*F)-D)+(1-(2*(ABSF↑2))))
30 P=G+(C↑2/(16*H))*(8*(J↑2)*(I-1)*(V)+
   (H*J))
32 R=(1-C)*O*P
35 A1=J*(C+ABSC↑2)):A3=(((D*C)/(2*H))*(ABS
   H↑2)+(2*(ABSJ↑2)))
37 A2=((I*ABSC↑2)/4)*(H*F-(5*J)+(4*(ABSJ↑2))
   *(1/(TAN(180*J)/π)))
38 Q=A1-A3+A2:U=(180/π)*((SINE*COSA*COSB)
   /H)*(Q)+E
40 W=(SINB*COSA)-(COSU*SINA*COSB):X=SINU*COSB
41 A4=ATN(X/W)
42 IF A4<0 LET Y=A4+180:GOTO47
45 Y=A4
47 IFE<0LET Z=Y+I80:GOTO51
50 Z=Y
51 IF Z=0ANDL<KTHENLETZ=180
```

```
52 Z=INT((Z*100)+.5)/100:PRINT"TC=";Z:R=INT
   ((R*100)+.5)/100:PRINT"DIST=";R
53 END
60 PRINT"DEP IDENT= ";Z$:PRINT"LAT=";K:
   PRINT"LNG=";M
62 PRINT"DES IDENT= ";Y$:PRINT"LAT=";L:
   PRINT"LNG=";N:RETURN
100 ZD=INT(XD*.01):ZM=INT(XD-(ZD*100)):
    ZS=XD-INTXD
105 ZT=((ZD*3600)+(ZM*60)+(ZS*6))/3600
110 RETURN
```

To use this program enter the information requested. Enter the Latitude and Longitude published in Flight Information Publications without the W or N or dashes; use all zeros found in these publications.

EXAMPLE:

N38-39.6 W090-38.7
is entered
3839.6 09038.7

Although this program is long, it is the only one I've found that allows you to go due North or due South and not produce an error message (divide by zero). This program cannot be used for courses that go to or pass over the North Pole. When using Latitudes north of 80 degrees, you will sometimes get a negative number; if this happens just disregard the sign.

The next formula and program are shorter but sacrifice some accuracy. However, they are accurate enough for most Great Circle Route calculations required by pilots.

GREAT CIRCLE FORMULA #2

When the terrestrial latitudes and longitudes of two points are given, this formula allows you to calculate the shortest distance between the two points and the azimuth at the start point.

$$D = 60 \ \text{Cos}^{-1} \left\{ \text{Sin}N_1 \text{Sin}N_2 + \text{Cos}N_1 \text{Cos}N_2 \text{Cos}(E_2 - E_1) \right\}$$

$$\theta_i = \text{Cos}^{-1} \left\{ \frac{\text{Sin}N_2 - \text{Sin}N_1 \text{Cos} \left(\frac{D}{60}\right)}{\text{Sin} \left(\frac{D}{60}\right) \text{Cos}N_1} \right\}$$

$$\theta = \begin{cases} \theta_i & \text{if } \text{Sin}(E_2 - E_1) \geq 0 \\ 360 - \theta_i & \text{otherwise} \end{cases}$$

WHERE:

N_1 = North Latitude Start

E_1 = East Longitude Start

N_2 = North Latitude End

E_2 = East Longitude End

D = Distance between points

Θ_i = Azimuth at Start Point(changes inflight)

North latitudes and east longitudes should be expressed by positive numbers, and south latitudes and west longitudes by negative numbers.

GREAT CIRCLE NAVIGATION
BASIC LANGUAGE PROGRAM #2

```
10 INPUT"N1=";N,"W1=";E,"N2=";M,"W2=";F
15 E=E*(-1):F=F*(-1)
20 A=N:GOSUB500:N=A:A=M:GOSUB500:M=A:A=E
   :GOSUB500:E=A
30 A=F:GOSUB500:F=A
40 D=60*ACS(SINN*SINM+COSN*COSM*COS(F-E))
50 S=ACS((SINM-SINN*COS(D/60))/SIN(D/60)/COSN)
60 IF SIN(F-E)<0THEN LET S=360-S
70 S=INT((S*100)+.5)/100:D=INT((D*100)+.5)/100
80 PRINT"DIST=";D:PRINT"T.C.=";S
500 AD=INT(A*.01):AM=INT(A-(AD*100)):AS=A-INTA
505 A=((AD*3600)+(AM*60)+(AS*6))/3600:RETURN
```

To use this program, enter the Latitude and Longitude as presented in Flight Information Publications but do not use Letters or spaces or dashes. Use all zeros.

EXAMPLE:

N38-39.6 W090-38.7
entered as
3839.6 09038.7

WHERE:

N1 = Latitude Departure
W1 = Longitude Departure

N2 = Latitude Destination
W2 = Longitude Destination

Courses to or over the North Pole or courses that are due North or due South will generate error conditions in this program (divide by zero).

South Latitudes and East Longitudes must be entered as negative numbers.

The following formula and program may be used to determine the Latitude where the course line crosses a given Longitude. More than one intermediate point could be used thereby dividing the trip into a number of legs. Doing this will produce a number of courses that will result in a more nearly Great Circle Course. The greater the number of intermediate points the more closely a Great Circle Route can be flown.

It is possible to use the body of the Basic Language Program as a subroutine in conjunction with the preceding Great Circle programs to establish a number of legs along a desired route.

This in fact is what I have done with the programs I use. I find that 4 legs are adequate for total route distances under 1500 miles. For the higher latitudes you may want to reduce this distance. Such reduction is not necessary for operations in the continental U.S.

Given the latitude and longitude of two points on the globe and an intermediate longitude, it is possible to calculate the the latitude corresponding to the intersection of the great circle route and the intermediate longitude.

GREAT CIRCLE PLOTTING FORMULA:

$$Lat_1 = \tan^{-1}\left[\frac{A-B}{\sin(Lng_D - Lng_S)}\right]$$

Where:

$A = (\tan Lat_D \cos Lng_S - \tan Lat_S \cos Lng_D)$
$\times \sin Lng_1$

$B = (\tan Lat_D \sin Lng_S - \tan Lat_S \sin Lng_D)$
$\times \cos Lng_I$

Lat_S = Latitude Start Point

Lng_S = Longitude Start Point

Lat_D = Latitude Destination

Lng_D = Longitude Destination

Lng_I = Longitude Intermediate Point

GREAT CIRCLE PLOTTING
BASIC LANGUAGE PROGRAM:

```
10 INPUT"DEP LAT=";LS,"DEP LNG=";MS,"DES LAT="
   ;LD,"DES LNG=";MD,"ILNG=";L1
12 L2=L1
15 Z=LS:GOSUB100:LS=Z:Z=MS:GOSUB100:MS=Z:Z=LD
   :GOSUB100:LD=Z:Z=MD:GOSUB100:MD=Z
17 Z=L1:GOSUB100:L1=Z
20 A=((TAN LD*COS MS)-(TAN LS*COS MD))*SIN L1
30 B=((TAN LD*SIN MS)-(TAN LS*SIN MD))*COS L1
40 L=ATN(ABS((A-B)/(SIN(MD-MS))))
45 DG=INT L:DM=(L-DG)*.6:DD=INT((((DG+DM)*100)
   *10)+.05)/10
50 PRINT"LAT=";DD:PRINT"LNG=";L2
60 END
100 AD=INT(Z*.01):AM=INT(Z-(AD*100)):AS=Z-INTZ
105 Z=((AD*3600)+(AM*60)+(AS*6))/3600:RETURN
```

To use this program enter the information requested. Enter the information without using the W or N or dashes. Southern Latitudes and Eastern Longitudes are entered as negative numbers.

EXAMPLE:

N38-39.6 W090-38.7
entered as
3839.6 09038.7

Note: No leg may pass half way around the Earth, and lines of longitude may not be plotted (due North or due South).

The words "dead reckoning" seem to spark a twinge of fear or at least anxiety in the hearts of most pilots. I think this is because we tend to have memories of FAA written exams, wind triangles, and the use of "whiz wheel confusers".

For the past few years I have been using a simple set of tables that apply to the speed of the airplanes I currently fly. These tables enable me to plan a flight very quickly and with amazing accuracy. They can be used enroute to maintain a "how-goes-it" check on the flight path and to check the accuracy of the winds aloft forecast.

I have provided three tables to determine the estimated crab angle and groundspeed when the true course, expected true airspeed and forcast winds are known.

To use the tables simply go to table one and determine the headwind or tailwind and crosswind components, then go to table two to determine the crab angle required for your expected true airspeed. If the crab angle exceeds 10 degrees then you must use table three to get the groundspeed loss due to crabbing. That's all there is to it! No E-6B, no wind triangle, no button pushing!

The tables may seem long and intimidating; the reason for this is that I have provided figures for a great number of airspeeds and wind conditions. You can find the values for your airplane and use a highlighter pen to make them easy to find.

The tables begin on page 129

WIND SPEED	ANGLE BETWEEN TRUE COURSE AND WIND DIRECTION																			
	0°		10°		20°		30°		40°		50°		60°		70°		80°		90°	
	X	HT	X	HT	X	HT	X	HT	X	HT	X	HT	X	HT	X	HT	X	HT	X	HT
10	10	0	10	2	9	3	9	5	8	6	6	8	5	9	3	9	2	10	0	10
20	20	0	20	3	19	7	17	10	15	13	13	15	10	17	7	19	3	20	0	20
30	30	0	30	5	28	10	26	15	23	19	19	23	15	26	10	28	5	30	0	30
40	40	0	39	7	38	14	35	20	31	26	26	31	20	35	14	38	7	39	0	40
50	50	0	49	9	47	17	43	25	38	32	32	38	25	43	17	47	9	49	0	50
60	60	0	59	10	56	21	52	30	46	39	39	46	30	52	21	56	10	59	0	70
70	70	0	69	12	66	24	61	35	54	45	45	54	35	61	24	66	12	69	0	70
80	80	0	79	14	75	27	69	40	61	51	51	61	40	69	27	75	14	79	0	80
90	90	0	89	16	85	31	78	45	69	58	58	69	78	45	31	85	16	89	0	90
100	100	0	98	17	94	34	87	50	77	64	64	77	87	50	94	34	17	98	0	100

This table provides the crosswind and headwind or tailwind components for various wind angles and speeds.
The X column is the crosswind component and the HT cloumn is the headwind or tailwind component. For wind speeds greater than 100 add components. Use knots or mph. Do not mix knots and mph.

TABLE 1

WIND COMPONENT TABLE

Figure 9-1

TRUE AIRSPEED	CROSSWIND COMPONENT FROM TABLE 1																			
	5	10	15	20	25	30	35	40	45	50	55	60	65	70	75	80	85	90	95	100
80	4°	7°	11°	14°	18°	22°	26°	30°	34°	39°	43°	49°	54°	61°	70°	90°	90°	90°	90°	90°
100	3	6	9	12	14	17	20	24	27	30	33	37	41	44	49	53	58	64	72	90
120	2	5	7	10	12	14	17	19	22	25	27	30	33	36	39	42	45	49	52	56
140	2	4	6	8	10	12	14	17	19	21	23	25	28	30	32	35	37	40	43	45
160	2	4	5	7	9	11	13	15	16	18	20	22	24	26	28	30	32	34	36	39
180	2	3	5	6	8	10	11	13	14	16	18	19	21	23	25	26	28	30	32	34
200	1	3	4	6	7	9	10	12	13	14	16	17	19	20	22	24	25	27	28	30
220	1	3	4	5	7	8	9	10	12	13	14	16	17	19	20	21	23	24	26	27
240	1	2	4	5	6	7	8	10	11	12	13	14	16	17	18	19	21	22	23	25
260	1	2	3	4	6	7	8	9	10	11	12	13	14	16	17	18	19	20	21	22
280	1	2	3	4	5	6	7	8	9	10	11	12	13	14	16	17	18	19	20	21

Don't mix speeds. If using Knots TAS use Knots wind speed.
For wind components over 100 add the wind correction angles.
NOTE: If wind correction angle greater than 10 see table 3.

TABLE 2
WIND CORRECTION ANGLE TABLE

Figure 9-2
See next page for remainder of this table.

CROSSWIND COMPONENT FROM TABLE 1

TRUE AIRSPEED	5	10	15	20	25	30	35	40	45	50	55	60	65	70	75	80	85	90	95	100
300	1°	2°	3°	4°	5°	6°	7°	8°	9°	10°	11°	12°	13°	13°	14°	15°	16°	17°	18°	19°
320	1	2	3	4	4	5	6	7	8	9	10	11	12	13	14	14	15	16	17	18
340	1	2	3	3	4	5	6	7	8	8	9	10	11	12	13	14	14	15	16	17
360	1	2	3	4	5	5	6	6	7	8	9	10	10	11	12	13	14	14	15	16
380	1	2	2	3	4	5	5	6	7	8	8	9	10	11	11	12	13	14	14	15
400	1	1	2	3	4	4	5	6	6	7	8	9	9	10	11	12	12	13	14	14
420	1	1	2	3	3	4	5	5	6	7	8	8	9	10	10	11	12	12	13	14
440	1	1	2	3	3	4	5	5	6	7	7	8	8	9	10	10	11	12	12	13
460	1	1	2	2	3	4	4	5	6	6	7	7	8	9	10	10	11	11	12	13
480	1	1	2	2	3	4	4	5	5	6	7	7	8	8	9	10	10	11	11	12
500	1	1	2	2	3	3	4	5	5	6	6	7	7	8	9	9	10	10	11	12

Don't mix speeds. If using Knots TAS use Knots wind speed.
Add wind correction angles for wind components over 100.
NOTE: If wind correction angle greater than 10 see table 3.

TABLE 2 (continued)
WIND CORRECTION ANGLE TABLE

Figure 9-2

	WIND CORRECTION ANGLE										
TAS	10°	15°	20°	25°	30°	35°	40°	45°	50°	55°	60°
80	1	3	5	7	11	14	19	23	29	34	40
100	2	3	6	9	13	18	23	29	36	43	50
120	2	4	7	11	16	22	28	35	43	51	60
140	2	5	8	13	19	25	33	41	50	60	70
160	2	5	10	15	21	29	37	47	57	68	80
180	3	6	11	17	24	33	42	53	64	77	90
200	3	7	12	19	27	36	47	59	71	85	100
220	3	7	13	21	29	40	51	64	79	94	110
240	4	8	14	22	32	43	56	70	86	102	120
260	4	9	16	24	35	47	61	76	93	111	130
280	4	10	17	26	38	51	66	82	100	119	140
300	5	10	18	28	40	54	70	88	107	128	150
320	5	10	19	30	43	58	75	94	114	136	160
340	5	12	21	32	46	61	80	100	121	145	170
360	5	12	22	34	48	65	84	105	129	154	180
380	6	13	23	36	51	69	89	111	136	162	190
400	6	14	24	37	54	72	94	117	143	171	200
420	6	14	25	39	56	76	98	123	150	179	210
440	7	15	27	41	59	80	103	129	157	188	220
460	7	16	28	43	62	83	108	135	164	196	230
480	7	16	29	45	64	87	112	141	171	205	240
500	7	16	29	45	64	87	112	141	171	205	240

This table gives the groundspeed LOSS due to crabbing into the wind when flying a given True Airspeed. REMEMBER, SUBTRACT THE VALUE FROM THIS TABLE FROM THE GROUNDSPEED DETERMINED BY USING TABLE 1.

TABLE 3
GROUNDSPEED LOSS DUE TO CRABBING

Figure 9-3

The formulas from which these tables were constructed are as follows:

FIND CROSSWIND COMPONENT:

XW = SIN WA x WS

FIND HEADWIND OR TAILWIND COMPONENT:

HT = COS WA x WS

FIND WIND CORRECTION ANGLE

WCA = ASN((SIN WA x WS) ÷ TAS)

FIND SPEED LOSS DUE TO CRABBING

SL = TAS - (COS WCA x TAS)

Where:

WA = Angle between true course and wind direction.

WS = Windspeed.

XW = Crosswind component.

HT = Headwind or tailwind component.

WCA = Wind correction angle.

TAS = True airspeed.

SL = Groundspeed loss due to crabbing.

TRUE HEADING AND GROUNDSPEED FORMULA

$$TH = TH_1 - \left[INT\left(\frac{TH_1}{360}\right) 360 \right]$$

$$GS = (V \cos W) + HT$$

Where:

$$TH_1 = TC - W$$

$$W = \operatorname{asn}\left(\frac{B \sin A}{V}\right)$$

$$HT = B \cos A$$

$$A = 180 + WD + \left|360 - TC\right|$$

B = Windspeed

WD = Wind direction

TC = True course

V = True airspeed

TH = True heading

GS = Groundspeed

This formula will give the true heading and groundspeed using forecast winds and expected true airspeed. The formula will work with either Knots or Statute miles per hour. Remember use all knots or all miles per hour, do not mix these values.

TRUE HEADING AND GROUNDSPEED BASIC LANGUAGE PROGRAM

```
10 INPUT"TC=";TC,"TAS=";V,"WND DIR=";WD,
   "WND SPD=";B
20 A=180+WD+360-TC:HT=B*COSA:W=ASN(B*SINA/V)
   :GS=V*COSW+HT:T=TC-W
30 TH=T-INT(T/360)*360:TH=INT((100*TH)+.5)
   /100:GS=INT((100*GS)+.5)/100
40 PRINT"TH=";TH;"    GS=";GS:END
```

To use this program just enter the information requested. The program will work with either Nautical or Statute values. However, don't mix Statute and Nautical values.

When you do your flight planning it is easy enough to determine the expected groundspeed and estimated time enroute based on enroute winds at the cruise altitude. But, there is always the problem of how much time to allow for the climb and descent. The following two rules of thumb may be of some help in solving this problem.

RULE OF THUMB #45 -- For light aircraft add one minute for each thousand feet to be climbed to the estimated time enroute.

RULE OF THUMB #46 -- For fast climbers (jets and most turboprops) figure the climb and descent segments using 85% of the expected enroute headwind or tailwind component.

If you are flying over hostile terrain or over water for long distances you should work a Radius of Action problem. This should be done in preflight planning and checked as soon as possible after establishing normal cruise flight. For preflight planning purposes use the information obtained from your dead reckoning calculations. Using the following two formulas you can determine both the Time and the Distance to the point of no return. The Distance to point of no return will give you the Distance from the start point to the point where the fuel required to return to the start point is the same as the fuel required to continue to the destination. The Time to the point of return will give you the time at which you must return to start point if the planned destination and alternate airports become unavailable for landing. These points may or may not coincide so figure both points and use the most restrictive of the two.

RADIUS OF ACTION

TIME TO POINT OF NO RETURN:

$$\frac{F_t}{H_{pnr}} = \frac{V_d + V_r}{V_r}$$

Where:

F_t = Total fuel in hours and decimals (less reserve fuel time).

V_d = Velocity (groundspeed) departure leg.

V_r = Velocity (groundspeed) return leg.

H_{pnr} = Hours and decimal hours to point of no return.

DISTANCE TO POINT OF NO RETURN:

$$\frac{D_t}{D_{pnr}} = \frac{V_{td} + V_r}{V_r}$$

Where:

D_t = Total distance.

V_d = Velocity (groundspeed) to destination.

V_r = Velocity (groundspeed) to return

D_{pnr} = Distance to point of no return.

RADIUS OF ACTION
BASIC LANGUAGE PROGRAM

```
10 INPUT"DEP G.S.=";D,"RETURN G.S.=";R,
   "FUEL(HRS.TENTHS)=";F,"TOTAL DIST.=";T
20 O=(D+R)/R:P=(F/O):A=T/O:P=INT((P*100)+.5)
   :A=INT((A*100)+.5)/100
30 PRINT"TIME TO PNR=";P:PRINT"DIST. TO PNR="
   A:END
```

This program gives the both the Time point of no return and the Distance point of no return. Enter the information requested. As in all other programs, if you use Knots then use Nautical miles.

The program will work with either Nautical or Statute values.

This information should be updated as soon as possible after establishing cruise flight to assure accuracy.

CALCULATION USING CR OR E-6B COMPUTERS

You can also use the calculator side of your E-6B to make the calculations. These are simple ratio problems. To do this either add the groundspeed departure leg or groundspeed to continue to the groundspeed to return (for planning simply double the expected true airspeed) find this value on the outer scale and place the groundspeed to return under the previous value and read the Time to PNR on the inner scale under the Fuel in hours, read the Distance to PNR under the Total Distance.

ENROUTE NAVIGATION

Now that you're enroute you find that the winds are not quite what you expected and you are a little off course. What to do? The answer is easy. Just consult the table in figure 9-4 to determine what heading to fly to parallel or to intercept your course.

To use the table all you need to do is first determine (by eyeball or otherwise) how far you are off course (DOC). Then, look in the distance column for a value near the distance flown (DF) and read the multiplier value in the multiplier column. Now, multiply the multiplier by the DOC; this will be the angle to parallel course (ATP).

If you want to parallel course all you have to do is turn this number of degrees into the wind. If you want to intercept your course at the destination or end of the leg all you have to do is find a value near your distance to fly (DTF) in the distance column and get that multiplier and multiply it by the DOC, add this answer to the ATP and this is the angle to intercept (ATI). Turn this number of degrees into the wind to intercept at the distance selected in the second part of the computation.

EXAMPLE:

DOC = 5 miles
DF = 50 miles
MULTIPLIER = 1.2 (from table 9-4)
ATP = 5 x 1.2 = 6.0 degrees
DTF = 200 miles
MULTIPLIER = .3 (from table 9-4)
ATI = .3 x 5 = 1.5 + 6.0 = 7.5 degrees

COURSE CORRECTION TABLE

DIST	MULTIPLIER	DIST	MULTIPLIER
1	60	210	.28
2	30	220	.27
3	20	230	.26
5	15	240	.25
6	12	250	.24
7	10	260	.23
8	8.6	270	.22
9	7.5	280	.21
10	6.7	290	.21
20	3.0	300	.20
30	2.0	310	.19
40	1.5	320	.19
50	1.2	330	.18
60	1.0	340	.18
70	.86	350	.17
80	.75	360	.17
90	.67	370	.16
100	.60	380	.16
110	.54	390	.15
120	.50	400	.15
130	.46	410	.15
140	.43	420	.14
150	.40	430	.14
160	.38	440	.14
170	.35	450	.13
180	.34	460	.13
190	.32	470	.13
200	.30	480	.12

ATP = Distance Multiplier x DOC

ATI = (Distance Multiplier x DOC) + ATP

Figure 9-4

ENROUTE WIND DIRECTION AND VELOCITY

USING TABLES TO FIND ENROUTE WIND

The tables in Figure 9-2 (pages 130 & 131) can be used to determine the approximate enroute wind vector. To do this find your true airspeed then move right to the column that is closest to your wind correction angle (the difference between heading and course) and read the crosswind component at the top of the table.

Once you have done this, find the difference between your true airspeed and groundspeed, this will be the headwind or tailwind component.

If the wind correction angle is greater than 10 degrees then use the table in Figure 9-3 (page 132) and subtract this value from the headwind or tailwind value.This will be the headwind or tailwind component.

Now, all you have to do is use the headwind and crosswind values and enter them into the graph in Figure 9-5 (on page 143). The value from this graph will be the enroute wind vector.

To determine the mangnetic wind direction use one of the computations below:

If wind is headwind from left - subtract the wind angle from your magnetic course.

If wind is headwind from right - add the wind angle to your magnetic course.

If wind is tailwind from left - add the wind angle to the reciprocal of your magnetic course.

If wind is tailwind from right - subtract the wind angle from the reciprocal of your magnetic course.

To get the true wind direction, correct wind direction for variation. Since you are correcting from magnetic to true you must ADD easterly or SUBTRACT westerly variation.

EXAMPLE:

TAS = 200
VAR = 5E
MH = 010
MC = 360
WCA = 10 degrees right
GS = 190

HEADWIND COMPONENT = 200 - 190 = 10

CORRECTED HEADWIND = 10 - 3 = 7 (Fig. 9-3)

CROSSWIND COMPONENT = 35 (Fig. 9-2)

WIND VECTOR = 78 degrees (Fig. 9-5)

WINDSPEED = 36 (Fig. 9-5)

MAGNETIC WIND = 360 + 78 = 78 degrees

TRUE WIND = 78 + 5 = 83 degrees

If you use knots for your computations the answer will be in knots. If you use miles per hour the answer will be in miles per hour. Don't mix these two values.

WIND COMPONENT GRAPH

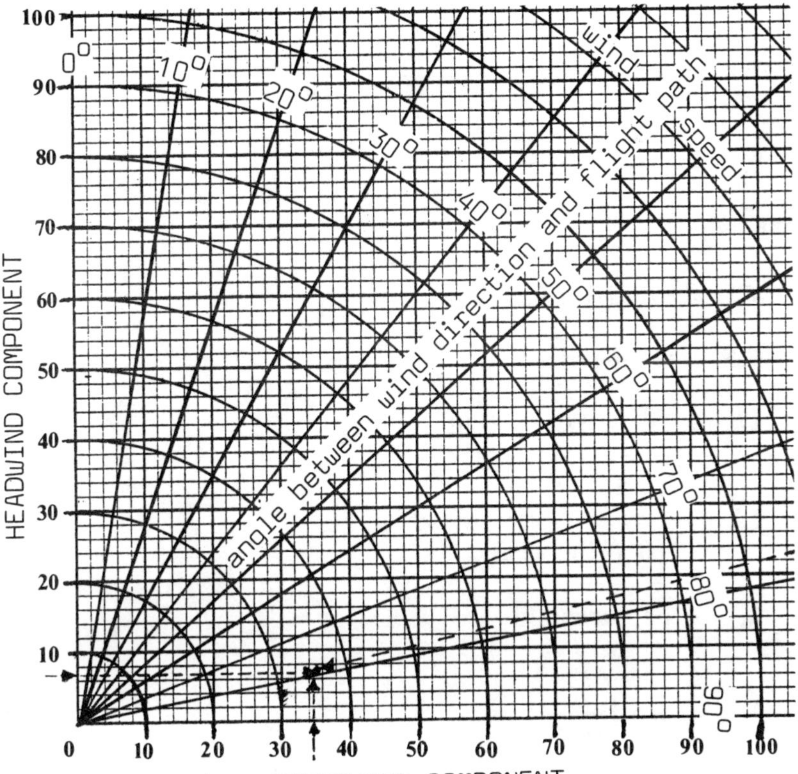

EXAMPLE:

CROSSWIND COMPONENT = 35

HEADWIND COMPONENT = 7

WIND ANGLE = 78°

WINDSPEED = 36

Figure 9-5

FORMULA FOR ENROUTE WIND:

$$WD = W_1 - \left[INT \left(\frac{W_1}{360} \right) 360 \right]$$

$$WS = \sqrt{HT^2 + C^2}$$

Where:

WD = wind direction

WS = windspeed

$W_1 = TC + W_2 + 180L'$

$$L' = \begin{cases} 1 \text{ if } GS > V \\ 0 \text{ otherwise} \end{cases}$$

V = true airspeed

$$W_2 = \operatorname{asn} \left(\frac{C}{WS} \right) L$$

$$L = \begin{cases} -1 \text{ if } V < GS \\ 0 \text{ if } V = GS \\ 1 \text{ otherwise} \end{cases}$$

$C = V \sin W$

$HT = (V \cos W) - GS$

$W = TH - TC$

GS = groundspeed

TH = true heading

TC = true course

ENROUTE WIND
BASIC LANGUAGE PROGRAM

```
10 INPUT"TRU CRS=";TC,"TRU HDG=";TH,"TAS=";
   V,"GRND SPD=";GS

20 W=TH-TC:HT=V*COSW-GS:C=V*SINW:WS=√((ABS
   HT↑2)+(ABSC↑2))

30 W2=ASN(C/WS)*SGN(V-GS):W1=TC+W2+180*(GS>V)
   :WD=W1-INT(W1/360)*360

40 WD=INT((100*WD)+.5)/100:WS=INT((100*WS)+.5
   )/100

50 PRINT"DIR=";WD;"    SPD=";WS:END
```

To use this program simply enter the information requested. Remember not to mix knots and mph. If you use knots the answer will be in knots. If you use miles per hour the answer will be in miles per hour.

USEFUL INFLIGHT RULES OF THUMB

RULE OF THUMB #47 -- The angle of bank required to make a standard rate of turn equals the TAS divided by 10, plus the constant 7; if the TAS is in knots or plus the constant 5; if the TAS is in MPH.

RULE OF THUMB #48 -- The lead angle required to roll out on a pre-selected heading in degrees equals:

One half the bank angle if standard rate turn.

One third the bank angle if half standard rate.

15 degrees if the bank angle is 45 degrees.

RULE OF THUMB #49 — When making timed turns divide the number of degrees to be turned by 3; this will be the time in seconds to make the turn when using standard rate turn. Multiply this figure by 2 if making half standard rate turn.

RULE OF THUMB #50 -- When making turns with the Magnetic Compass, the compass will lag behind the turn when on or turning to northerly headings. The compass will lead the turn when on or making turns to southerly headings. The amount of lead or lag will be equal to the Latitude over which the turn is being made.

RULE OF THUMB #51 -- When leveling from climb or descent begin the recovery by 10% of the climb or descent rate, or 50 feet; whichever is greater.

RULE OF THUMB #52 — The approximate inflight visibility is equal to one mile for each thousand feet altitude when you can just see the surface over the nose.

CHAPTER 10

ELECTRONIC NAVIGATION

CHAPTER CONTENTS

DME CORRECTION CALCULATIONS

The DME, while a great piece of electronic gear, is subject to a few minor errors. The following information will help you correct for these errors.

SLANT RANGE CORRECTION

This function corrects the slant range error. This error is not large but increases as altitude increases or distance decreases.

FORMULA:

$$\text{DISTANCE} = \frac{(\Delta H)^2 + \text{DIST}^2}{\text{DIST}}$$

Where:

ΔH = Altitude above DME station in feet
DIST = DME distance in feet (miles x 6076)

BASIC LANGUAGE PROGRAM

```
10 INPUT"IND. ALT.=";AI,"VOR ALT.=";AF,
   "DME DIST.=";HI

20 H=HI*6076:A=AI-AF:DA=√(H↑2-A↑2):D=DA/6076

30 D=INT((100*D)+.5)/100

40 PRINT"ACTUAL DIST.=";D:END
```

To use the program enter the information requested. The answer is the actual distance in nautical miles.

DME SPEED CORRECTION

The groundspeed given by a DME is only accurate when flying directly toward or away from the station. Using the following table you may correct the indicated groundspeed for flight paths that do not go directly to or from the station. To use the table just determine the angle between the radial you are crossing and your magnetic heading and multiply the indicated groundspeed by the multiplier for this angle. The accuracy decreases for angles over 60 degrees.

DME SPEED CORRECTION TABLE

ANGLE	MULTIPLIER	ANGLE	MULTIPLIER
5	1.00	35	1.22
10	1.02	40	1.30
15	1.04	45	1.41
20	1.06	50	1.56
25	1.10	55	1.74
30	1.15	60	2.00

Figure 10-1

EXAMPLE:

TAS = 200. ANGLE = 25. MULTIPLIER = 1.10

GROUNDSPEED = 200 x 1.10 = 220

ACTUAL GROUNDSPEED = 220 knots

DME SPEED CORRECTION FORMULA:

$$GS = \left| \frac{SPD_{dme}}{\cos(RAD - CRS)} \right|$$

Where:

RAD = Crossing Radial

CRS = Magnetic Heading

SPD_{dme} = Indicated Groundspeed

BASIC LANGUAGE PROGRAM

```
10 INPUT"IND. G.S.=";I,"RADIAL=";A,"HEADING=";C

20 GS=ABS(I/COS(A-C)):GS=INT((100*GS)+.5)/100

30 PRINT"G.S.=";GS:END
```

This program calculates the groundspeed from the DME speed indicator when the aircraft is not flying directly to or from a DME station.

To use the program, enter the information requested. The accuracy of the program suffers when the aircraft crosses the radials at an angle larger than 60 degrees.

TIME AND DISTANCE CHECKS

These checks, while seldom done in "real world" conditions, are sometimes required during flight training and check rides.

$$\text{Time To Station} = \frac{60 \times \text{Minutes Flown}}{\text{Degrees Change}}$$

$$\text{Dist. To Station} = \frac{\text{TAS} \times \text{Minutes Flown}}{\text{Degrees Change}}$$

RULE OF THUMB #53 -- A simple way to determine time to station is to use a 10 degree bearing change and divide the time in seconds by 10. The answer will be the time to the station in minutes.

These calculations will work with both NDB and VOR stations.

COURSE INTERCEPTION

RULE OF THUMB #54 -- To determine an intercept heading simply use any heading between the heading necessary to go to the station and the radial or bearing to be intercepted. Make the proper heading adjustment to intercept for inbound or outbound tracking. Works with both NDB and VOR stations.

RULE OF THUMB #55 -- When using an ADF, when the relative bearing increases the station is to the right of the aircraft. When the relative bearing decreases, the station is to the left of the aircraft.

RULE OF THUMB #56 -- When using the VOR, when the radials (FROM indication) increase the station is to the right of the aircraft. When radials decrease the station is to the left of the aircraft.

COURSE DISPLACEMENT OF AIRCRAFT

RULE OF THUMB #57 -- When using the VOR, aircraft displacement from course is approximately 200 feet per dot on the CDI per nautical mile.

RULE OF THUMB #58 -- One degree off course in 60 miles is approximately equal to one mile off course.

PILOT NOTES

CHAPTER 11

DESCENT AND APPROACH

CHAPTER CONTENTS

RULES OF THUMB

TABLES

FORMULAS

BASIC LANGUAGE PROGRAMS

ENROUTE DESCENTS AND CROSSING RESTRICTIONS

This section will discuss descent profiles to make crossing restrictions. It will try to answer the questions of when to start descent, where to start descent and what rate to use for the descent.
Several methods are presented here, so pick the one that best suits your needs.
Before getting into the heat of the subject matter there are a few helpful rules of thumb to consider.

RULE OF THUMB #59 -- The number of miles per minute is approximately equal to the Mach number times 10. Or, the Groundspeed divided by 60.

RULE OF THUMB #60 -- A 1° pitch change will give a descent rate approximately equal to your speed in miles per minute times 100, or by using rule #59, 1000 times your Mach number.

RULE OF THUMB #61 -- To figure the number of vertical feet per mile is represented by a descent gradient, simply multiply the gradient by 100. Thus, a 4° gradient represents a 400 foot vertical change per mile.

RATE OF DESCENT FORMULA

$$\text{RATE} = \frac{\text{CRUISE ALT - CROSSING ALT}}{\left(\dfrac{\text{MILES TO FIX}}{\text{(GROUNDSPEED/60)}}\right)}$$

This formula computes the rate of descent required to reach a fix at a specified altitude.

RATE OF DESCENT
BASIC LANGUAGE PROGRAM

```
10 INPUT"CRUISE ALT=";A,"XING=";C,
   "MI TO FIX=";M,"G.S.=";S

20 G=S/60:T=M/G:B=A-C:R=B/T

30 R=INT((100*R)+.5)/100:PRINT"RATE=";R:END
```

To use this program, enter the information requested. Use the estimated groundspeed to allow for speed loss during descent. Don't mix knots and mph.

DESCENT START POINT FORMULA

$$\text{DIST} = \frac{(\text{CRUISE ALT-CROSSING ALT}) \times \text{G.S.}}{\text{RATE} \times 60}$$

This formula computes the distance from a fix to begin descent at a predetermined rate to reach the fix at the desired altitude.

DESCENT START POINT
BASIC LANGUAGE PROGRAM

```
10 INPUT"R.O.D.=";D,"G.S.=";S,"CRUISE ALT=";A
   ,"XING ALT=";C

20 W=A-C:X=W/D:Y=S/60:Z=X*Y:Z=INT((100*Z)+.5)
   /100

30 PRINT"T.O.D.=";Z:END
```

To use the program, enter the information requested. Don't mix knots and mph.

APPROACH

DETERMINE DISTANCE AND TIME ON DME ARC

This is a very useful function to help determine when to expect interception of the final approach course when flying a DME arc.

TIME ON ARC FORMULA:

$$\text{TIME} = 60\left(\frac{((2\pi R)\ 360)D}{G}\right)$$

Where:

R = Radius of arc in nautical miles

D = Degrees (difference between entry radial and final course radial).

G = Estimated average groundspeed.

DISTANCE ON ARC FORMULA:

$$\text{MILES} = D\left(\frac{2\pi R}{360}\right)$$

Where:

R = Radius of arc in nautical miles.

D = Degrees (difference between entry radial and final course radial).

TIME AND DISTANCE ON DME ARC
BASIC LANGUAGE PROGRAM

```
10 INPUT"ARC RADIUS=";R,"EST G.S.=";G,
   "DGS ON ARC=";D

20 T=(((π*(R*2)/360)*D)/G

30 M=T*G:PRINT"MILES=";M

40 PRINT"TIME=";T*60:END
```

To use the program, enter the data requested. The groundspeed should be estimated to correct for wind effect while on arc, and is entered in knots.
The Distance will be in nautical miles and the time will be in minutes and decimal minutes.

APPROACH RATE OF DESCENT (3° GLIDESLOPE)

RULE OF THUMB #62 -- The rate of descent in feet per minute required to stay on a 3° glideslope equals the groundspeed in knots divided by 2. Add a zero to the quotient. Thus, if the groundspeed is 100 then the rate of descent will be 500 feet per minute.

See The table in Figure 11-1 for rates of descent for various glideslope angles and groundspeeds.

The program for enroute rate of descent may be used for these calculations by using the glideslope intercept altitude for Cruise Altitude and the DH or MDA for the Crossing Altitude.

DESCENT RATE TABLE

A rate of descent table is provided for use in planning and executing precision descents under known or approximate ground speed conditions. It will be especially useful for approaches when the localizer only is used for course guidance. A best speed, power, attitude combination can be programmed which will result in a stable glide rate and attitude favorable for executing a landing if minimums exist upon breakout. Care should always be exercised so that the minimum descent altitude and missed approach point are not exceeded.

ANGLE OF DESCENT (degrees and tenths)	GROUND SPEED (knots)										
	30	45	60	75	90	105	120	135	150	165	180
2.0	105	160	210	265	320	370	425	475	530	585	635
2.5	130	200	265	330	395	465	530	595	665	730	795
3.0	160	240	320	395	480	555	635	715	795	875	955
3.5	185	280	370	465	555	650	740	835	925	1020	1110
4.0	210	315	425	530	635	740	845	955	1060	1165	1270
4.5	240	355	475	595	715	835	955	1075	1190	1310	1430
5.0	265	395	530	660	795	925	1060	1190	1325	1455	1590
5.5	290	435	580	730	875	1020	1165	1310	1455	1600	1745
6.0	315	475	635	795	955	1110	1270	1430	1590	1745	1905
6.5	345	515	690	860	1030	1205	1375	1550	1720	1890	2065
7.0	370	555	740	925	1110	1295	1480	1665	1850	2035	2220
7.5	395	595	795	990	1190	1390	1585	1785	1985	2180	2380
8.0	425	635	845	1055	1270	1480	1690	1905	2115	2325	2540
8.5	450	675	900	1120	1345	1570	1795	2020	2245	2470	2695
9.0	475	715	950	1190	1425	1665	1900	2140	2375	2615	2855
9.5	500	750	1005	1255	1505	1755	2005	2255	2510	2760	3010
10.0	530	790	1055	1320	1585	1845	2110	2375	2640	2900	3165
10.5	555	830	1105	1385	1660	1940	2215	2490	2770	3045	3320
11.0	580	870	1160	1450	1740	2030	2320	2610	2900	3190	3480
11.5	605	910	1210	1515	1820	2120	2425	2725	3030	3335	3635
12.0	630	945	1260	1575	1890	2205	2520	2835	3150	3465	3780

Figure 11-1

PILOT NOTES

CHAPTER 12

LANDING PERFORMANCE

CHAPTER CONTENTS

RULES OF THUMB

Landing performance is not as greatly affected by atmospheric variations as takeoff performance because little engine power is required. So, I won't bore you with a lot of tables, graphs and formulas.
Instead, I will give you a number of really useful rules of thumb. I use many of these all the time and find them a great help.

LANDING AND APPROACH SPEED CALCULATIONS

RULE OF THUMB #63 -- Speed over the threshold (Vref) = 1.30% Vso (Vso is the bottom of the white arc).

RULE OF THUMB #64 -- Approach speed = Vref + 10 + 1/2 headwind + gust factor. To be reduced to Vref over threshold.

STOPPING DISTANCE CALCULATIONS

RULE OF THUMB #65 -- A 1% change in airspeed will cause a 2% change in stopping distance.

RULE OF THUMB #66 -- A 10% change in landing weight will cause a 10% change in stopping distance.

RULE OF THUMB #67 -- A 1000 foot change in field elevation will cause a 4% change in stopping distance. The greater the altitude, the greater the distance required. Due to increase in TAS.

RULE OF THUMB #68 — A 15°F deviation from standard temperature will cause a 4% change in stopping distance. The greater the termerature, the greater the stopping distance. Due to increase in TAS.

RULE OF THUMB #69 -- A headwind which is 10% of the landing speed will reduce the landing distance approximately 19%, while a tailwind which is 10% of the landing speed will increase the landing distance 21%.

RULE OF THUMB #70 -- Crossing the threshold 50 feet too high increases the landing distance 25%.

RULE OF THUMB #71 -- A 1° reduction in glide angle increases the landing distance by 13%.

RULE OF THUMB #72 -- A wet runway can increase the stopping distance 25% to 50%.

RULE OF THUMB #73 -- An icy runway can increase stopping distance by 100% or more.

These calculations are based on sea level standard day figures from the flight manual. Each will either add to or subtract from previous applications of these rules.

CHAPTER 13

CONVERSIONS
AND
MISCELLANEOUS CALCULATIONS

CHAPTER CONTENTS

PILOT NOTES

CONVERSIONS

VOLUME

MULTIPLY	BY	TO GET
U.S GALLONS	3.7854	LITERS
	.83268	IMPERIAL GALLONS
IMPERIAL GAL.	4.546	LITERS
	1.2009	U.S. GALLONS
LITERS	.26417	U.S. GALLONS
	.21997	IMPERIAL GALLONS

VOLUME CONVERSIONS BASIC LANGUAGE PROGRAM

```
10 INPUT"UGAL(1),IGAL(2),LITERS(3)";A

12 A=INTA

15 IFA<1ORA>3THEN10

20 ON A GOTO100,200,300

100 INPUT"U.S. GALLONS=";G

110 I=G*.83268:R=I:GOSUB400:I=R:L=G*3.7854
    :R=L:GOSUB400:L=R

120 PRINT"IGAL=";I;" LIT=";L:END

200 INPUT"IMP. GALLONS=";I

210 G=I*1.2009:R=G:GOSUB400:G=R:L=I*4.546
    :R=L:GOSUB400:L=R
```

```
220 PRINT"US GAL=";G;" LIT=";L:END

300 INPUT"LITERS=";L

310 G=L*.26417:R=G:GOSUB400:G=R:I=L*.21997
    :R=I:GOSUB400:I=R

320 PRINT"US GAL=";G;" IGAL="I:END

400 R=INT((R*10)+.05)/10:RETURN
```

To use this program select the number for the value you want to convert **from.** Then enter the values requested. The answer will be the other two values. For example, if you enter the number 2, the program will convert the value entered in Imperial gallons to U.S. gallons and Liters.

This is a dandy program for those of you who fly outside the U.S. from time to time.

FUEL WEIGHT

JET A at 60°F / 15°C = 6.75 lbs/gal.

AVGAS at 60°F / 15°C = 6.0 lbs/gal.

See the Fuel Weight conversion tables in Figure 13-1 and Figure 13-2.

Also, see the Density Variation of Aviation Fuel in Figure 13-3.

FUEL WEIGHT TABLE

JET A - JP5 at 60°F / 15°C

UNITS	U.S. GAL. 6.75 lbs/gal	IMP. GAL. 8.15 lbs/gal	LITERS 1.79 lbs/gal
1	6.75	8.15	1.79
5	33.75	40.75	9.00
10	67.50	81.50	17.90
15	101.25	122.25	26.90
20	135.00	163.00	35.80
25	168.75	203.75	44.80
50	337.50	407.50	89.50
100	675.00	815.00	179.00
150	1,012.50	1,222.50	268.50
200	1,350.00	1,630.00	358.00
250	1,687.50	2,037.50	447.50
300	2,025.00	2,445.00	537.00
350	2,362.00	2,852.00	626.50
400	2,700.00	3,260.00	716.00
450	3,037.50	3,667.50	805.50
500	3,375.00	4,075.00	895.00
550	3,712.50	4,482.50	984.50
600	4,050.00	4,890.00	1,074.00

Figure 13-1

FUEL WEIGHT TABLE (cont'd)

AVIATION GASOLINE at 60°F / 15°C

UNITS	U.S. GAL. 6.00 lbs/gal	IMP. GAL. 7.20 lbs/gal	LITERS 1.79 lbs/gal
1	6.00	7.20	1.58
2	12.00	14.40	3.17
3	18.00	21.60	4.75
4	24.00	28.80	6.34
5	30.00	36.00	7.92
6	36.00	43.20	9.50
7	42.00	50.40	11.09
8	48.00	57.60	12.67
9	54.00	64.80	14.26
10	60.00	72.00	15.84
20	120.00	144.00	31.68
30	180.00	216.00	47.52
40	240.00	288.00	63.36
50	300.00	360.00	79.20
60	360.00	432.00	95.04
70	420.00	504.00	110.80
80	480.00	576.00	126.72
90	540.00	648.00	142.56
100	600.00	720.00	158.40

Figure 13-2

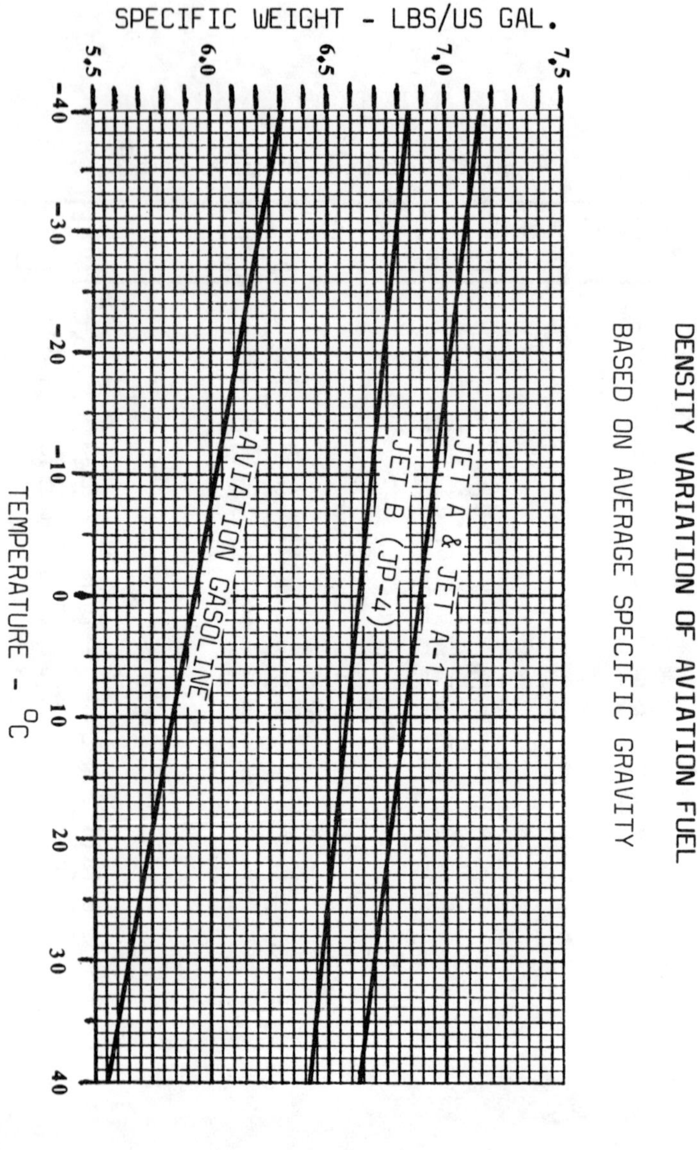
DENSITY VARIATION OF AVIATION FUEL
BASED ON AVERAGE SPECIFIC GRAVITY
SPECIFIC WEIGHT - LBS/US GAL.
5,5
6,0
6,5
7,0
7,5
-40
-30
-20
-10
0
10
20
30
40
TEMPERATURE - °C
JET A & JET A-1
JET B (JP-4)
AVIATION GASOLINE

Figure 13-3

FUEL POUNDS TO GALLONS CONVERSION
BASIC LANGUAGE PROGRAM

```
10 INPUT"JET A(1),  AV GAS(2)";A

20 A=INTA

25 INPUT"TEMP C=";T,"FUEL LBS. NEEDED=";P

30 IF A<1ORA>2 THEN10

40 ON A GOTO 100,200

100 F=((T*.0068*-1)+6.89):G=P/F:F=INT((F*10)
    +.05)/10:G=INT((G*10)+.05)/10

110 PRINT"POUNDS PER GAL=";F

120 PRINT"GALLONS=";G:END

200 F=((T*.009375*-1)+5.93):G=P/F:F=INT((F*10)
    +.05)/10:G=INT((G*10)+.05)/10

210 PRINT"POUNDS PER GAL=";F

220 PRINT"GALLONS=";G:END
```

This program will give the fuel in U.S. gallons required to give the fuel weight input. The program also corrects for temperature. To use the program enter either 1 or 2 and the information requested. If fuel is stored above ground use ambient temperature; if underground storage use 15°C.

DISTANCE AND SPEED

MULTIPLY	BY	TO GET
NAUTICAL MILES	6076.1	FEET
	1.1508	STATUTE MILES
	1.8520	KILOMETERS
STATUTE MILES	5280.0	FEET
	.86898	NAUTICAL MILES
	1.6094	KILOMETERS
KILOMETERS	3280.8	FEET
	.53995	NAUTICAL MILES
	.62136	STATUTE MILES

If you use Nautical miles per hour, Statute miles per hour, etc., you will get the answer in converted units per hour.

DISTANCE AND SPEED
BASIC LANGUAGE PROGRAM

```
10 INPUT"NMI(1), SMI(2), KM(3)";A

20 A=INTA

30 IFA<1ORA>3THEN10

40 ON A GOTO 100,200,300

100 INPUT"NAUTICAL MI=";N:S=N*1.1508:S=INT
    ((S*10)+.05)/10

110 K=N*1.8520:K=INT((10*K)+.05)/10

120 PRINT"SMI=";S;"  KM=";K:END
```

```
200 INPUT"STATUTE MILES=";S:N=S*.86898:N=INT
    ((N*10)+.05)/10

210 K=S*1.6094:K=INT((K*10)+.05)/10

220 PRINT"NMI=";N;"   KM=";K:END

300 INPUT"KILOMETERS=";K:N=K*.53995:N=INT
    ((N*10)+.05)/10

310 S=K*.62136:S=INT((S*10)+.05)/10

320 PRINT"NMI=";N;"   SMI=";S:END
```

To use the program enter the number of the value **from** which you want to convert. Enter the information requested. The answer will be other two values.
Example to convert Statute Miles, enter 2, then enter the information requested; Statute Miles will be converted to Nautical Miles and Kilometers.

TEMPERATURE CONVERSION

$$F^{O} = 9/5(C^{O}+40)-40$$

$$C^{O} = 5/9(F^{O}+40)-40$$

$$R^{O} = F^{O}+459.67$$

$$K^{O} = K^{O}+273.15$$

OR

$$C^{O} = (F^{O}-32)/1.8$$

$$F^{O} = 1.8C^{O}+32$$

CELSIUS, FAHRENHEIT CONVERSION
BASIC LANGUAGE PROGRAM

```
10 INPUT"C TO F (1)     F TO C (2)";A:A=INTA

20 IFA<1ORA>2THEN10

30 ON A GOTO 100,200

100 INPUT"TEMP C="C:F=(1.8*C)+32:F=INT((100
    *F)+.5)/100

110 GOTO300

200 INPUT"TEMP F=":C=(F-32)/1.8:C=INT((C*100
    *F)+.5)/100
210 GOTO300

300 PRINT" C=";C;"     F=";F:END
```

To use the program enter the number for the conversion to be made, then enter the requested information.

STANDARD ATMOSPHERE TABLE

ALT. FT.	TEMP. °F	TEMP. °C	PRESS. IN. Hg.
0	59.0	15.0	29.92
1000	55.4	13.0	28.86
2000	51.9	11.0	27.82
3000	48.3	9.1	26.82
4000	44.7	7.1	25.84
5000	41.2	5.1	24.90
6000	37.6	3.1	23.98
7000	34.0	1.1	23.09
8000	30.5	-0.8	22.22
9000	26.9	-2.8	21.39
10000	23.3	-4.8	20.58
11000	19.8	-6.8	19.79
12000	16.2	-8.8	19.03
13000	12.6	-10.8	18.29
14000	9.1	-12.7	17.58
15000	5.5	-14.7	16.89
16000	1.9	-16.7	16.22
17000	-1.6	-18.7	15.57
18000	-5.2	-20.7	14.94
19000	-8.8	-22.6	14.34
20000	-12.3	-24.6	13.75
21000	-15.9	-26.6	13.18
22000	-19.5	-28.6	12.64
23000	-23.0	-30.6	12.11
24000	-26.6	-32.5	11.60
25000	-30.2	-34.5	11.10

Figure 13-4

STANDARD ATMOSPHERE TABLE

ALT. FT.	TEMP. °F	TEMP. °C	PRESS. IN. Hg.
26000	-33.7	-36.5	10.63
27000	-37.3	-38.5	10.17
28000	-40.9	-40.5	9.725
29000	-44.4	-42.5	9.297
30000	-48.0	-44.4	8.885
31000	-51.6	-46.4	8.488
32000	-55.1	-48.4	8.106
33000	-58.7	-50.4	7.737
34000	-62.2	-52.4	7.382
35000	-65.8	-54.3	7.041
36089	-69.7	-56.5	6.683
37000	-69.7	-56.5	6.397
38000	-69.7	-56.5	6.097
39000	-69.7	-56.5	5.811
40000	-69.7	-56.5	5.538
41000	-69.7	-56.5	5.278
42000	-69.7	-56.5	5.030
43000	-69.7	-56.5	4.794
44000	-69.7	-56.5	4.569
45000	-69.7	-56.5	4.355
46000	-69.7	-56.5	4.151
47000	-69.7	-56.5	3.956
48000	-69.7	-56.5	3.770
49000	-69.7	-56.5	3.593
50000	-69.7	-56.5	3.425

Figure 13-4

GLOSSARY

AIRSPEED TERMINOLOGY

IAS--Indicated Airspeed is the speed shown on the airspeed indicator.

CAS--Calibrated Airspeed is indicated airspeed corrected for position and instrument error.

TAS--True Airspeed is calibrated airspeed corrected for temperature, pressure, and compressability effects.

GS--Ground Speed, though not an airspeed, is calculated directly from TAS if true wind direction and speed are known.

M--Mach Number is the ratio of true airspeed to the speed of sound.

Vmc--Minimum Control Speed is the minimum speed at which an airplane is controllable. With twin-engine airplanes this speed is the minimum speed with one engine inoperative and the other engine developing takeoff power. Maximum bank angle is 5 degrees.

Vmcg--Ground Minimum Control Speed.

Va--Maneuvering speed is the maximum speed at which application of full available aerodynamic control will not overstress the airplane.

Vmo/Mmo--Maximum Operating Limit Speed. Either CAS (Vmo) or MACH (Mmo).

Vfe--Maximum Flap Extended Speed is the highest speed allowed with wing flaps in a prescribed position.

Vle--Maximum Landing Gear Extended Speed is the highest speed at which an aircraft may be safely flown with the landing gear extended.

Vlo--Maximum Landing Gear Operating Speed is the maximum speed the landing gear may be extended or retracted (i.e. intransit).

Vlof--Lift Off Speed.

Vr--Rotation Speed.

Vs--Power Off Stalling Speed or the Minimum Steady Flight Speed at which the airplane is controllable.

Vso--The Stalling Speed or the Minimum Steady Flight Speed in the landing configuration.

Vx--Best Angle of Climb Speed.

Vxse--Single Engine Best Angle of Climb Speed for twin-engine airplanes.

Vy--Best Rate of Climb Speed

Vyse--Single Engine Best Rate of Climb Speed for twin-engine airplanes.

V1--Takeoff Decision Speed.

V2--Initial Climb Speed (usually engine out climb speed for Transport Category Aircraft. Similar to Vxse.).

Vref--Approach Reference Speed is equal to 1.3 Vso for the actual aircraft gross weight.

METEOROLOGICAL TERMINOLOGY

INDICATED PRESSURE ALTITUDE--The value displayed on the Altimeter when the barometric scale has been set to 29.92 inches of mercury (1013 millibars).

PRESSURE ALTITUDE--Indicated altitude corrected for Altimeter error.

DENSITY ALTITUDE--Pressure altitude corrected for non-standard temperature.

ALTIMETER SETTING--Barometric Pressure corrected to sea level.

OAT--Outside Air Temperature is the free air static temperature, obtained either from ground meteorological sources or from inflight temperature indicators and adjusted for compressibility effects.

ISA--International Standard Atmosphere in which:

(1) the air is a perfect dry gas;

(2) the temperature at sea level is 59 degrees Fahrenheit, 15 degrees Celsius;

(3) the pressure at sea level is 29.92 inches mercury;

(4) the temperature gradient from sea level to the altitude at which the temperature is -69.7 degrees Fahrenheit is -0.003566 degrees Fahrenheit per foot and zero above that altitude.

GLOSSARY

POWER TERMINLOLGY

TAKE OFF POWER--Is the maximum power rating and limited to a maximum of 5 minutes operation. Use of this rating should be limited to normal take-off operations and emergency situations.

MAXIMUM CONTINUOUS POWER--Is the highest power rating not limited by time. Use of this rating is intended for emergency situations at the discretion of the pilot.

CRUISE CLIMB POWER--Is the maximum power approved for normal climb.

MAXIMUM CRUISE POWER--Is the highest power rating for cruise and is not time limited. Observe airframe limitations.

PERFORMANCE TERMIMOLOGY

ACCELERATE-GO--The distance required to accelerate to V1, experience an engine failure, continue accelerating to Vlof then climb and accelerate in order to achieve V2 at 35 feet above the runway.

ACCELERATE-STOP--The distance required to accelerate to V1 and stop.

AGL--Above Ground Level.

BEST ANGLE OF CLIMB SPEED--The Best Angle of Climb Speed is the airspeed which delivers the qreatest gain of altitude in the shortest possible horizontal distance with gear and flaps retracted, or in specified position.

BEST RATE OF CLIMB SPEED--The Best Rate of Climb Speed is the airspeed which delivers the greatest gain in altitude in the shortest possible time with gear and flaps retracted, or in specified position.

CLIMB GRADIENT--The ratio of change in height during a portion of a climb, to the horizontal distance traveled in the same time period.

NET CLIMB GRADIENT--The gradient of climb with the flaps in the take-off position, and the gear retracted. Net indicates that the actual gradient of climb has been reduced by 0.8% to allow for pilot technique and turbulence.

DEMONSTRATED CROSSWIND--The maximum 90 degree crosswind component for which adequate control of the aircraft during take-off and landing was actually demonstrated during certification.

TAKE-OFF FLIGHT PATH--The minimum gradient of climb required to clear obstacles in excess of 35 feet, measured horizontally from reference zero and vertically at the altitude above the runway. Reference zero is the point where the airplane has reached 35 feet above the runway as determined from applicable aircraft graphs.

CLEARWAY--An area beyond the runway not less that 300 feet on either side of the extended runway centerline, at an elevation no higher than the elevation of the runway end, clear of all fixed obstacles, and under the control of the airport authority.

TAKE-OFF WEIGHT--Weight of the airplane at lift-off.

LANDING WEIGHT--The actual weight of the airplane at touchdown.

WEIGHT AND BALANCE TERMINOLOGY

APPROVED LOADING ENVELOPE--Those combinations of airplane weight and center of gravity which define the limits beyond which loading is not approved.

ARM--The distance from the center of gravity of an object to a line about which moments are to be computed.

CENTER OF GRAVITY--A point at which the weight of an object may be considered concentrated for weight and balance purposes.

DATUM--A vertical plane perpendicular to the airplane longitudinal axis from which fore and aft (usually aft) measurements are made for weight and balance purposes.

MOMENT--A measure of the rotational tendency of a weight, about a specified line, mathematically equal to the product of the weight and the arm.

STATION--The longitudinal distance from some point to the zero datum or zero fuselage station.

TARE--The apparent weight which may be indicated by a scales before any load is applied.

BASIC EMPTY WEIGHT--The airplane weight including unusable fuel, full oil, and full operating fluids.

EMPTY WEIGHT--The airplane weight with fixed ballast, unusable fuel, engine oil, engine coolant, hydraulic fluid, and other items required by regulatory standards.

BASIC OPERATING WEIGHT--Is the Empty Weight plus required crew, and any other items normally carried in the aircraft (navigation charts, galley supplies, etc.).

MAXIMUM WEIGHT--The greatest weight allowed by design, structural, performance or other limitations.

RAMP WEIGHT--The gross weight of the airplane before engine start. Includes Take-Off weight plus fuel allowed for start, taxi, run-up and take-off ground roll.

UNUSABLE FUEL--The fuel remaining after consumption of all usable fuel.

USABLE FUEL--The amount of total fuel which is available for consumption.

USEFUL LOAD--The difference between the airplane ramp weight and the basic empty weight.

PAYLOAD--The difference between the Basic Operating Weight plus the Required Fuel and the Maximum Weight Allowed or the difference between Zero Fuel Weight and the Basic Operating Weight; whichever is smaller.

ZERO FUEL WEIGHT--The airplane ramp weight minus the weight of fuel on board. Stated another way, this is the maximum weight allowed with no fuel in the wing tanks. On airplanes with fuselage fuel tanks this weight is known as the Maximum Wing Bending Weight. In any case it is a structural limit, and once the weight build-up has reached the specified limit, all new weights must be fuel in the wing tanks.

ENGINE OIL--That portion of the engine oil which can be drained from the engine.

PILOT NOTES

APPENDIX A

FLIGHT PLAN SEQUENCE

1. TYPE
   VFR, IFR, DVFR
2. AIRCRAFT IDENT.
3. AIRCRAFT TYPE/
   SPECIAL EQUIPMENT
4. TRUE AIRSPEED
5. DEPARTURE POINT
6. DEPARTURE TIME
7. CRUISE ALTITUDE
8. ROUTE OF FLIGHT
9. DESTINATION
10. ESTIMATED TIME ENROUTE
11. REMARKS
12. FUEL ON BOARD
13. ALTERNATE
14. PILOTS NAME, ETC.
15. NUMBER ON BOARD
16. COLOR

EQUIPMENT SUFFIX

X = No Transponder
T = Transponder No Altitude Reporting
U = Transponder With Altitude Reporting
D = DME No Transponder
B = Transponder & DME, No Altitude Reporting
A = Transponder & DME, With Altitude Reporting
M = Tacan Only
N = Tacan & Transponder, No Altitude Reporting
P = Tacan & Transponder, With Altitude Reporting
C = Rnav & Transponder, No Altitude Reporting
R = Rnav & Transponder, With Altitude Reporting
W = Rnav, No Transponder

Library of Progress

VOLUME 39, NUMBER 8

by Sari March Schnepf

Air Progress *Magazine*

IFR Pocket Simulator Procedures

What do you say about a book claiming to be a "pocket simulator?" Well, you pick it up and see what it's all about. Indeed, in light of the trend towards using the comparatively less expensive flight simulators (as compared to the real thing) the author Henry H. Culver has come up with the idea of an instrument training book which will provide the same training as a simulator or an airplane at only $19.95. Actually the title doesn't tell the whole story. There isn't a more thorough book to cover the expensive work that schools put their students through, and Culver believes that much of this work could be done by studying his manual. And I would hardly disagree.

The author covers every in-flight radio navigational procedure and attitude flying technique that most approved schools require in their instrument flying course curriculum. Culver claims that instead of having to pay $15 or $45 per hour to be introduced to these procedures while seated in a flight simulator or aircraft cockpit, you can now learn them economically in the privacy of your own home and prepare yourself at your own pace for subsequent lessons. This way you'll be able to apply your assimilated knowledge and earn the rating with minimum expense.

Claims may be great, you say, but how does the book live up to its designs? Well, the format is the most interesting part and it definitely distinguishes it from its counterparts — Instrument Flying by Robert Taylor and the FAA's Instrument Flying Handbook. Each chapter is broken down into specific areas and each area has a procedure and subsequent steps that delineate the procedure and make that area of knowledge easier to understand. Then the author tests you on your knowledge of that particular area. For example: In the chapter on Radio Orientation, he breaks it down into VOR, ADF, and triangulation with a DF steer. Then, ADF for instance, is broken down into three parts: ADF homing, ADF station passage and ADF position identification. These parts in turn are separated into procedures which are explained by steps. At the end of the ADF part, you are confronted with sample problems, to make sure you know how you're understanding.

As far as content, IFR Pocket Simulator Procedures examines Radio Orientation, Course Interceptions, Maintaining Course, Intersection Identification as well as Holding Patterns, Instrument Approaches and DME arcs. There's an entire section on attitude flying, which explores the four fundamentals aerodynamic forces and flight control coordination.

The information is complete and precise. The procedures are accurate and easy to understand, and the quizzes are great methods of testing how much you understand. In living up to the expectations as an instrument course supplement, the book includes an entire flying course syllabus, complete with lesson plans, objectives and completion standards.

To round out the program and make the book entirely acceptable for studying for the written exam, Culver incorporates sections on the fundamentals of radio and instrument operation, plus a detailed description of approach charts and plates, and finally a chapter on the "Illusions of flight." He artfully ends the book in a glossary of aeronautical terms to further clarify any misunderstandings and make the course easier still.

In assessing Culver's objective, I'd have to say that this book is foremost in supplementing any instrument course. The author leaves no stone unturned—the content is excellent. But its redeeming value lies in its format. The information is organized and clear.

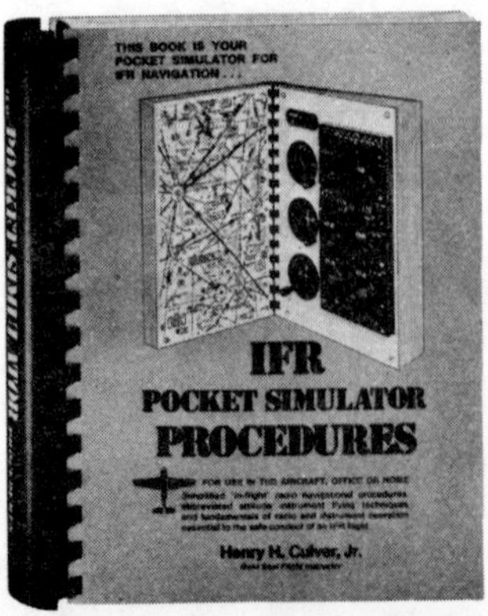

INTRODUCING

Dial-a-Panel™

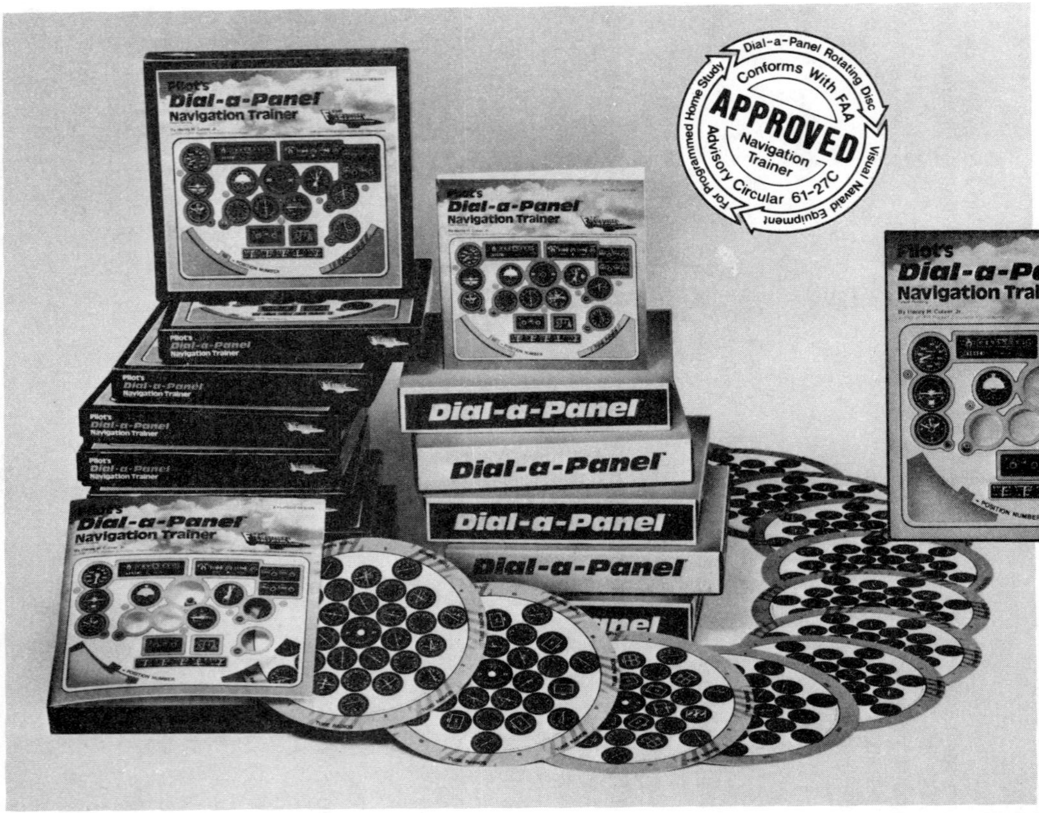

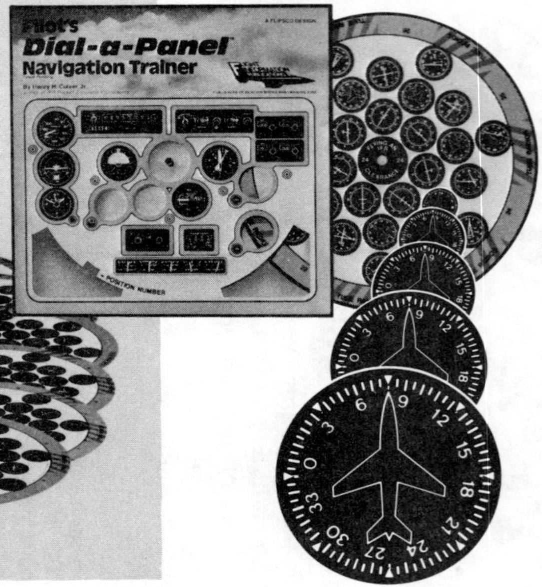

Actual size D.G.

aviation's lowest-cost simulator . . .